SOCIAL THEORY
POLITICAL PRAC

CW00521440

SOCIAL THEORY
AND
POLITICAL PRACTICE

Wolfson College Lectures 1981

EDITED BY

CHRISTOPHER LLOYD

CLARENDON PRESS · OXFORD
1983

Oxford University Press, Walton Street, Oxford OX2 6DP

London Glasgow New York Toronto
Delhi Bombay Calcutta Madras Karachi
Kuala Lumpur Singapore Hong Kong Tokyo
Nairobi Dar es Salaam Cape Town
Melbourne Auckland
and associates in
Beirut Berlin Ibadan Mexico City Nicosia

Published in the United States by
Oxford University Press, New York

British Library Cataloguing in Publication Data
Social theory and political practice.
1. Political sociology
I. Lloyd, Christopher
301   JA76
ISBN 0–19–827447–5
ISBN 0–19–827448–3 Pbk

Library of Congress Cataloging in Publication Data
Main entry under title:
Social theory and political practice.
(Wolfson College lectures; 1981)
Includes index.
Contents: Reflections on social theory and
political practice/by Ralf Dahrendorf —
Social theory and politics in the history of
social theory/by Tom Bottomore — Political
theory and practice/by Charles Taylor — [etc.]
1. Sociology — Addresses, essays, lectures.
2. Political sociology — Addresses, essays, lectures.
3. Ideology — Addresses, essays, lectures. I. Lloyd,
Christopher, 1950-   .   II. Series
HM33.S62 1983   301   82–14555
ISBN 0–19–827447–5
ISBN 0–19–827448–3 (pbk.)

Typeset by Hope Services
Printed in Great Britain
at the University Press, Oxford
by Eric Buckley
Printer to the University

PREFACE

Wolfson College, a recently founded graduates' college in the University of Oxford, has held every year since 1970 a series of public lectures on topics which are thought to be of general interest to the community and for which leading scholars are invited. This book contains the enlarged and revised versions of seven of the eight 1981 lectures which were delivered in the Hall of the College in Hilary Term. (Professor Stuart Hall, who delivered the eighth lecture, was unable to revise it in time for publication due to pressure of work.) As organizer and editor I thank the President, Sir Henry Fisher, the Fellows, and the students of the College for their encouragement and support; and in particular I thank Dr Wlodzimierz Brus for his assistance with the organization and Gillian Moore for her efficient and cheerful handling of the many administrative and secretarial tasks involved with the planning, organization, and publishing of the series. Most importantly, thanks are due to the lecturers themselves for taking part and making the series a stimulating and memorable addition to the Oxford intellectual feast.

November 1981   CHRISTOPHER LLOYD

Contents

Notes on Contributors   viii

 Editor's Introduction   1
   by Christopher Lloyd

1. Reflections on Social Theory and   25
   Political Practice
     by Ralf Dahrendorf

2. Social Theory and Politics in the History   39
   of Social Theory
     by Tom Bottomore

3. Political Theory and Practice   61
     by Charles Taylor

4. Accounts, Actions, and Values: 87
   Objectivity of Social Science
     by Amartya Sen

5. Social Theory, Social Understanding,   109
   and Political Action
     by John Dunn

6. The Collapse of Consensus: Ideology in   137
   British Politics
     by David Marquand

7. Marxism and Communism   157
     by Wlodzimierz Brus

Index of Names   181

Notes on Contributors

RALF DAHRENDORF: born in Germany and educated at the Universities of Hamburg and London. He has taught and researched at universities in Germany and the United States, was for a short period a member of the West German government, and for four years was a member of the Commission of the European Economic Community. Since 1974 he has been Director of the London School of Economics and Political Science.

TOM BOTTOMORE: born in England and educated at London University. He has held positions in universities in Britain, Canada, the United States, and India and been Secretary and President of the International Sociological Association. Since 1968 he has been Professor of Sociology at Sussex University.

CHARLES TAYLOR: born in Canada and educated at McGill and Oxford Universities. He has held positions in universities in Canada, Britain, France, and the United States, and until 1981 was Chichele Professor of Social and Political Theory at Oxford University. He is now Professor of Political Science at McGill University, Canada.

AMARTYA SEN: born in India and educated at the Universities of Calcutta and Cambridge. He has worked in universities in India, the United States, and Britain, was a Professor of Economics at Delhi University, and the London School of Economics, and is now Drummond Professor of Political Economy at Oxford University.

JOHN DUNN: born in England and educated at Cambridge University. He has held positions in universities in Britain, the United States, Ghana, and Canada. Since 1977 he has been Reader in Politics at Cambridge University.

DAVID MARQUAND: born in England and educated at Oxford University. He has held posts in universities in Britain and the United States, been a newspaper journalist, a Labour Member of Parliament from 1966 to 1977, and chief adviser to the Secretary-General of the European Commission. He is now Professor of Contemporary History and Politics at Salford University.

WLODZIMIERZ BRUS: born in Poland and educated at the Universities of Saratov and Warsaw. Prior to 1968 he was Professor of Political Economy at Warsaw University and Director of the Research Bureau of the Polish Planning Commission. Since coming to Britain in 1972 he has held posts at Glasgow and Oxford Universities and is now University Lecturer in Modern Russian and East European Studies and a Fellow of Wolfson College, Oxford University.

CHRISTOPHER LLOYD: born in Australia and educated at the University of New England (New South Wales) and at Sussex and Oxford Universities. He has been a newspaper journalist and has taught at New England. He is currently doing research at Wolfson College, Oxford.

Introduction

Christopher Lloyd

I

The relationship of social theory to political practice is an old
but still important and contentious question for the social
sciences. There has been a complicated and shifting inter-
relationship between theorists and political actors extending
back at least as far as the early eighteenth century, when
systematic sociological enquiry was first begun, and back
very much further if political philosophers and political
economists are seen as embodying a concern to enquire into
the nature of political action from a perspective of theorizing
about the nature of society in general. While the theory/
politics relationship has never been ignored by social thinkers
or politicians there has been in recent years an increase in
interest in the question due, perhaps, to the related influences
of economic and social crises and the breakdown of the
dominant mode of functionalist and positivist social and
political theorizing of the post-war capitalist world. Such
political and theoretical consensus as did exist in the liberal
democracies has considerably weakened, especially in Britain
where conflicting theories and ideologies have come to play
a much more overt and powerful role in politics. Even in
some totalitarian countries, despite the hegemony of official
ideologies and repressive regimes that have stifled discussion,
there has been lately some debate about the nature of society
and politics, such as in Poland and Brazil; but, sadly, in others
debate has become more restricted, as in Czechoslovakia
and Argentina.

This series of Wolfson College Lectures was planned and
organized in view of the new ferment over the theory/politics
relationship. We hoped at least to open up the issue a little by
having contributions from philosophical, historical, and

practical-political points of view. Obviously, a book of
lectures such as this cannot be a systematic introduction to
the subject, but what it lacks in comprehensive treatment and
coherence of perspective it gains in terms of juxtaposition of
different arguments and a demonstration that the issues are
not easily settled. The issues are complicated and should not
be oversimplified. Too often political discussion is based
upon blind prejudice and fuelled by demogogy. I hope this
book, like the public lectures, will be a small contribution
by the College towards political and theoretical education,
as well as a contribution to the defence of the social sciences
in a time of attack by showing their continuing relevance
to fundamentally important philosophical and practical
questions.

II

While the following contributors do raise, between them,
most of the central aspects of the theory/politics relationship, it is not done, on the whole, in a systematic manner,
and at times a certain degree of understanding of the field
by the reader is assumed. My aim in this introduction* is to
outline briefly what I see as some of the main underlying
issues in order that the papers and their arguments may be
better located by someone not having a good knowledge of
the field.

   In general by 'social theory' we mean here any form of
abstract theorizing about the nature of human society, economy, and social action; and by 'political practice' we mean
that kind of action which is directed towards using and
influencing governmental and state power. Now, both terms
are of course contentious. We can perhaps fairly easily recognize an explicitly developed and presented social theory,
which has the aim of abstracting in some way from the 'given'
reality of social life in order to accentuate or uncover or explain certain aspects, regularities, hidden tendencies, and

---

*I thank Michael Carrithers and Linley Lloyd for very helpful comments upon
earlier versions of this introduction.

influences. But some thinkers would go further and include all thinking and discussion about the nature of society, by anyone who has for whatever reason an interest in understanding how his or her society and political system work. This would obviously imply that almost everybody is at least an amateur social theorist.

Similarly, in modern liberal-democratic societies, political activity cannot, perhaps, be separated neatly from other kinds of behaviour since many actions have political effects. The specifications of the political domain and political action have long been a problem for theorists and philosophers. Should politics be seen primarily, say, as a set of techniques for achieving the particular social goals of particular groups, or perhaps as a set of ideas and practices which have the effect of legitimating the implementation of such goals? Or is it necessary that there be a public space for disputation about society and political goals — a forum for ideas — before there can be said to be politics?

The specifications of what constitute social theory and political practice are thus themselves theoretical issues which are raised in general in the following essays.

Systematic social theory can be viewed as being not only a means for the explanation and understanding of past and present politics but also as a potential guide, or indeed an imperative, to future political action. This view then raises philosophical questions about the possible connections between the three realms of social scientific theory and knowledge, commonsense understandings and values, and social reality. These connections centre upon the question of whether there can be shared objective social knowledge as a basis for social and political action. The desire for such knowledge is a very old one, originating primarily in the shadow of the successful natural sciences of the seventeenth and eighteenth centuries. Ever since then a central question for the philosophy of social enquiry has been whether it should attempt to be methodologically similar to certain natural sciences, which were taken to be exemplary objective discourses.

Traditionally there have been three broad alternative

responses to the question of the possibility of objective
social knowledge. One argues that it is possible but only
if the general (supposedly rational) explanatory orientation
of natural science is adopted as its method. This is the
broadly *positivist* response deriving chiefly from the work of
the early nineteenth-century thinkers Auguste Comte and
J. S. Mill, and of which there are two main sub-species,
depending upon whether the conception held of natural
science explanation is empiricist or realist. The empiricist
account argues that the basis of natural science is the empiricist
theory of knowledge. This holds that only those statements
which are or can be empirically verified can be considered as
true knowledge; that there should be a strict distinction
between theories and observations; that the task of objective
scientific enquiry is to seek for the true empirically-knowable
causes of phenomena, as revealed by constant correlations
of events; and that metaphysical ideas and judgements of
value have no place in science. There is an internal debate
within empiricist-positivism over whether the logic of theory
formulation, inference, and testing should be inductive or
deductive. The realist account of scientific method often rejects induction in favour of a deductive structure of theories
and inference, and sees the task of science as searching for the
real, but often hidden or non-observable, causal mechanisms
of experimental and natural phenomena.

A second response is to deny the possibility of objective
social knowledge broadly on the grounds of a fundamental
difference between the objects of enquiry of natural and
social science. Society, it is argued, is not a fixed, determinate,
external structure that impinges upon human consciousness
and action in some regular, knowable, and predictable way.
Rather, humans with their subjective intentions project
themselves into the world in complex, changing, and context-
bound ways, and only interpretive understanding, by people
who share their subjective pretheoretical meanings, is possible.
Any sharp separation of theories, facts, and values is then
ruled out. This account of social understanding sees it as a
process which links culturally-derived values, meanings, and

perceptions with individual experiences in a circular, self-reinforcing, self-validating manner, in order to arrive at authentic understanding. There cannot be then an objective source of truth or verification of interpretations. (This interpretive approach is often called *hermeneutical* after the Greek God Hermes, the God of speech, science, and so on, who acted as messenger and interpreter for the other Gods.)

A third response is to argue that a relativistic picture of enquiry should be generalized to include all the sciences to some extent. That is to say, all empirical enquiry, whether about society or nature, takes place within specific intellectual and cultural *traditions* and involves subjective, value-laden judgements, such that a dichotomy of objective and relative knowledge is an over-simplification. All enquiry, it is sometimes said, is conducted from particular points of view, using particular non-rational pre-scientific understandings, concepts, and values, and these must be recognized by philosophers as being not only present but indispensable. Nevertheless, rational explanation within particular traditions may still be possible.

Now it is true that these general philosophical positions may not be adopted completely, with all their ramifications, by any particular thinker, but they do underlie discussions about attempts to use theory both to explain political behaviour and as a guide to political interventions. Most political theories and practices contain conceptions of the good society and of the legitimacy of particular programmes of action designed to achieve it, and these conceptions in turn involve theories of how to change or engineer society. So it seems undeniable that political action, however we conceive it in general terms, must at least tacitly involve philosophical ideas about how we can come to know and understand the nature of society and politics. But the precise relationship that philosophy has with theory, commonsense understanding, and political practice, is certainly open to dispute, and as an illustration of this I want to consider briefly some aspects of an important debate that went on largely in Germany in the 1960s amongst the proponents, firstly, of a hermeneutical approach to society and politics; secondly, of an objectivist and piecemeal-engineering approach; and thirdly, of a dialectical-critical approach. I hope in this way that not only the question of the importance of underlying philosophical orientations to social knowledge will become a little clearer, but that the contours of the general social theory/political practice relationship will be illuminated. Several of the following essays do in fact deal with aspects of this debate but not in such a way as to make its outline clear.

III

One of the most important and lucid interpreters and developers of hermeneutical theory and philosophy in modern times has been the German philosopher Hans-Georg Gadamer, whose major work *Truth and Method* (1960)[1] has strongly influenced subsequent discussion. It took as its starting point the interpretation of art, and indeed, modern hermeneutical philosophy of social science has been considerably influenced by work in aesthetics, as well as biblical interpretation, and psychoanalysis — all areas in which the development and explication of interpretive theories have been central. Of course the German tradition of interpretive *Geisteswissenschaft* (science of mind or culture) goes back to the early nineteenth century and indirectly to Vico in the early eighteenth. The modern tradition traces its roots directly from Wilhelm Dilthey's work in the latter part of last century on the philosophy of history and social science.[2]

Gadamer has been concerned with the general problem of how *understanding* is possible and, a little more specifically,

[1] Hans-Georg Gadamer, *Truth and Method*, English translation of the 2nd (1965) edition, Steed and Ward, London, 1975.
[2] For discussions of hermeneutics in general see Richard E. Palmer, *Hermeneutics*, North western University Press, Evanston, 1969; Paul Rabinow and William M. Sullivan (eds.), *Interpretive Social Science*, University of California Press, Berkeley, 1979; and Josef Bleicher, *Contemporary Hermeneutics*, Routledge and Kegan Paul, London, 1980.

with what he has called (in a later essay that summarizes his philosophy) 'the central question of the modern age', viz: 'how our natural view of the world — the experience of the world that we have as we simply live our lives — is related to the unassailable and anonymous authority that confronts us in the pronouncements of science.'[3] He criticized the way in which 'natural' unreflective aesthetic and historical forms of consciousness had become alienated through striving for *objective* judgements about art and history; and hoped to show that 'it is not so much our judgements as it is our prejudices that constitute our being'.[4] The concept of *prejudice* had to be rehabilitated:

prejudices are not necessarily unjustified and erroneous, so that they inevitably distort the truth. In fact, the historicity of our existence entails that prejudices, in the literal sense of the word, constitute the initial directedness of our whole ability to experience. Prejudices are biases of our openness to the world. They are simply conditions whereby what we encounter says something to us.[5]

Thus people are open to the world — in a state of questioning or 'hermeneutical conditionedness' — and this is a universal state, for the scientist as much as the artist, or for anyone. But the sciences hide this state behind abstractions which serve to conceal many possibilities for questioning. Nevertheless, even in the sciences, as in life, 'no assertion is possible that cannot be understood as an answer to a question, and assertions can only be understood in this way'.[6] It is the asking of new questions, not scientific methodology, that leads to new knowledge, and this is the hermeneutical element of all experience, including science. It is this element which joins science and life experience together and reveals, in Gadamer's view, the fundamental level of the 'linguistic constitution of the world'.[7] Coming to know and understand the world in all its historical particularities is a process of constituting the world through language. We experience and

[3] H-G. Gadamer, 'The Universality of the Hermeneutical Problem', reprinted in Bleicher, op. cit., p. 128.   [4] Ibid., p. 133.
[5] Ibid., p. 133.   [6] Ibid., p. 135.   [7] Ibid., p. 136.

come to understand the world through the medium of
our pre-existing language and our linguistic-based cultural
tradition. Our consciousness of the world is affected by
history and fulfilled by language. Thus:

> There is always a world already interpreted, already organised in its
> basic relations, into which steps experience as something new, repressing
> what has led our expectations and undergoing reorganisation itself in
> the upheaval. Misunderstanding and strangeness are not the first factors,
> so that avoiding misunderstanding cannot be regarded as the specific
> task of hermeneutics. Just the reverse is the case. Only the support of
> familiar understanding makes possible the venture into the alien, the
> lifting up of something out of the alien, and thus the broadening and
> enrichment of our own experience of the world.[8]

In an earlier essay Gadamer considered at length the
problem of historical consciousness, a form of consciousness
which he wished to separate sharply from scientific knowledge
on the grounds that science is concerned with regularities and
generalities whereas historical knowledge should be of the
singularity and uniqueness of entities.[9] The form of historical
understanding had to be, he believed, that of the hermeneutical
circle. The historian tries to establish in a circularly reinforcing
manner the relationship between a whole and its parts, i.e.
between a socio-cultural totality and the lives, actions, and
products of individuals within it. In the act of interpretation
of a text or text analogue (such as an action or utterance) the
starting place is the whole formed by the subjectivity of the
author of the text, and that whole can be understood only by
a person who shares the cultural tradition of the author and
is thus able to mediate between the text and its meaning. The
coherence of the text and its context is *anticipated* by the
interpreter and that guides the interpretation in an immanent
manner. The original condition of a hermeneutical understanding is thus a *shared* or comprehensible reference to
things in themselves.[10] A hermeneutical interpretation must
distinguish the false from the true prejudices which illuminate

---

[8] Ibid., p. 138-9.

[9] H-G. Gadamer, 'The Problem of Historical Consciousness', reprinted in
Rabinow and Sullivan, op. cit., p. 116.   [10] Ibid., p. 155.

the text. *Objective history* is ultimately impossible, in Gadamer's opinion, because such a historian forgets the historicity
of his own position and seeks a non-historical (i.e. non-
hermeneutical) truth outside himself — he postulates the
historical process as a thing in itself, in the way that hermeneutics does, but then idealizes this thing as objective rather
than culturally relative.[11]

At the 1961 meeting of the German Sociological Association
an attempt was made to initiate a discussion between the
defenders of the hermeneutical conception of history and
social science and the proponents of an objectivist and
positivist attitude. The impetus for the debate came from a
growing realization by philosophers from both the pre-
dominatly Anglo-American analytical/empiricist tradition
and the Germanic idealist/hermeneutical tradition that there
was a gulf of understanding and attitude between them on
the question of the nature of the human studies and their
relationship to natural science, to 'commonsense' practical
reasoning, and to political practice.[12] Karl Popper, a leading
member of the analytical tradition (and emigrant from
inter-war Austria), was invited to address the meeting on the
logic of the social sciences. He had written widely on the
philosophy of science and the relationships of scientific
method and understanding to the solving of social problems.
In earlier works, notably *The Logic of Scientific Discovery*
(first published in German in 1934)[13] he had defended
a conception of science as being   essentially critical and
evolutionary, seeing it as typically employing a deductive
procedure (rather than inductive) to produce theories which
could then be *empirically* tested. A scientific statement was
in his view one which could be potentially empirically *falsified,*
although it could never be inductively proven completely

[11] Ibid., p. 157.
[12] For the background to the debate and a valuable discussion of its significance
see the Introduction by David Frisby to the English translation of T. W. Adorno
et al., The Positivist Dispute in German Sociology, Heinemann, London, 1976,
which contains the contributions to the debate up to 1970.
[13] Karl Popper, *The Logic of Scientific Discovery,* Hutchinson, London, 1959.

true, since he believed induction was fundamentally logically flawed as a method of inferring general truths. Knowledge for him was always provisional and must be made the object of criticism and constant testing.

At the 1961 conference he presented his main thesis in the following form:

(a) The method of the social sciences, like that of the natural sciences, consists in trying out tentative solutions to certain problems: the problems from which our investigations start, and those which turn up during the investigation.

Solutions are proposed and criticized. If a proposed solution is not open to pertinent criticism, then it is excluded as unscientific, although perhaps only temporarily.

(b) If the attempted solution is open to pertinent criticism, then we attempt to refute it; for all criticism consists of attempts at refutation.

(c) If an attempted solution is refuted through our criticism we make another attempt.

(d) If it withstands criticism, we accept it temporarily; and we accept it, above all, as worthy of being further discussed and criticized.

(e) Thus the method of science is one of tentative attempts to solve our problems; by conjectures which are controlled by severe criticism. It is a consciously critical development of the method of 'trial and error'.

(f) The so-called objectivity of science lies in the objectivity of the critical method. This means, above all, that no theory is beyond attack by criticism; and further, that the main instrument of logical criticism — the logical contradiction — is objective.[14]

Allied to this basic conception of Popper's were several other important ideas. Firstly, while he accepted that no science — natural or social — could be separated from the extra-scientific interests and values that influenced it, he wished to try to draw a distinction between those values and the values of scientific enquiry itself. No scientist could be free from cultural values, but he should strive to uphold the scientific values of truth, relevance, simplicity, and so on, attempting to separate the two realms of value.[15]

Secondly, it is thus the value commitment and task of pure science, including social science, to search for objectively true

[14] K. Popper, 'The Logic of the Social Sciences', in Adorno, op. cit., pp. 89–90.
[15] Ibid., pp. 96–8.

causal explanations. The basic logical schema for this must always consist in a deductive inference from theories and initial conditions. If these premises are true then logically the conclusion must be true, or if the conclusion is false then at least one of the premises must be false. Thus this standard of truthful causal explanation should become, in Popper's view, the regulative principle of a critical scientific procedure:

> The important methodological idea that *we can learn from our mistakes* cannot be understood without the regulative idea of truth: any mistake simply consists in a failure to live up to the standard of objective truth, which is our regulative idea. We term a proposition 'true' if it corresponds to the facts, or if things are as described by the proposition. This is what is called the absolute or objective concept of truth which each of us constantly uses. The successful rehabilitation of this absolute concept of truth is one of the most important results of modern logic.[16]

As this remark indicates, he believed the concept of truth to have been undermined by relativistic ideologies, the sorts of systems of ideas that he had combatted in earlier books, chiefly under the label of the 'sociology of knowledge', and which he had come to the German Sociological Conference to combat on its own ground, in the forms of idealist, dialectical, and critical theories.

Thirdly, social science should conform to this methodological prescription as much as natural science. A purely objective social science was possible, he believed, if it employed the method of objective understanding, in other words *situational* logic, so that it avoided all subjective or psychological ideas. Human action was to be explained by a deduction from knowledge of the situation of action, and objective understanding arose from the realization that action was appropriate to its situation. Rather strangely, perhaps, Popper included in his definition of the situation such seemingly subjective elements as wishes, motives, memories, and associations; as well as the physical world, and the social world of people and their institutions (the

---

[16] Ibid., p. 99.

largely unintended consequences of past actions) which influence action.[17]

Finally, Popper's critique of induction had been generalized, especially in his earlier books *The Open Society and its Enemies* and *The Poverty of Historicism*, into a critique of all attempts at social prediction and revolutionary politics and a defence of piecemeal social engineering. History has no intrinsic meaning, he argued, but it could be given a meaning through making a conscious decision to do so.[18] Historical facts and decisions about meaning had to be rigorously separated. This was the failure of 'historicism', which asserted, he said, that history did have an inherent direction and meaning and if we grasped it we could attune our lives accordingly. Historicism was a kind of fatalism, for Popper,[19] and was based upon the logically false procedure of inductively generalizing to establish the laws of historical evolution of social wholes, from which predictions about the future could be made.[20] Against this he proposed for social science a piecemeal technological purpose which rested upon an instrumentalist view of social theory. The task of the social engineer was to 'design social institutions, and to reconstruct and run those already in existence' in accordance with certain decided ends.[21] The piecemeal engineer never aims to remodel the whole of society in the way that a Utopian revolutionary does. One of the basic differences between the two approaches was that:

While the piecemeal engineer can attack his problem with an open mind as to the scope of the reform, the wholist cannot do this; for he has decided beforehand that a complete reconstruction is possible and necessary. This fact has far-reaching consequences. It prejudices the Utopianist against certain sociological hypotheses which state limits to institutional control . . . By a rejection *a priori* of such hypotheses, the Utopian approach violates the principles of scientific method. On

[17] Ibid., pp. 102–4.

[18] K. Popper, *The Open Society and Its Enemies,* 2 vols., Routledge and Kegan Paul, London, 5th ed., 1966; vol. 2, p. 269.

[19] Ibid., p. 279, and K. Popper, *The Poverty of Historicism*, Routledge and Kegan Paul, London, 1959, p. 51.   [20] Ibid., pp. 41–5.

[21] Ibid., pp. 64–5.

the other hand, problems connected with the uncertainty of the human factor must force the Utopianist, whether he likes it or not, to try to control the human factor by institutional means, and to extend his programme so as to embrace not only the transformation of Society, according to plan, but also the transformation of man . . .

It seems to escape the well-meaning Utopianist that this programme implies an admission of failure, even before he launches it. For it substitutes for his demand that we build a new society, fit for men and women to live in, the demand that we 'mould' these men and women to fit into his new society. This, clearly, removes any possibility of testing the success or failure of the new society. For those who do not like living in it only admit thereby that they are not yet fit to live in it; that their 'human impulses' need further 'organising'. But without the possibility of tests, any claim that a 'scientific' method is being employed evaporates. The wholistic approach is incompatible with a truly scientific attitude.[22]

Insofar as Popper defended this package of notions, which coalesced around the empirical testing of theories, the fact/ value distinction within science, the unification of natural and social scientific methods and the rejection of wholism, he can be considered as a kind of positivist (a label which he nevertheless rejected on the grounds that he believed positivism adopted an inductive basis for scientific unity).

The leading antagonist of positivism in the debate that followed the 1961 Conference was Jürgen Habermas, the chief inheritor of what has become known as the Frankfurt School of social thought, which developed from the 1920s.[23] This School was one of the most influential currents of Marxian thought to emerge in Europe after the Great War and it developed themes derived from the more Hegelian and dialectical elements in Marxism as well as from other traditions of Germanic philosophy. Its chief project, perhaps, has been to develop a critique of the alienation its members see at the heart of capitalist culture with its scientific-technological rationality.

From the 1960s onwards, Habermas has attempted to

---

[22] *Ibid.*, pp. 69–70.
[23] For an introduction to the work of the School see David Held, *Introduction to Critical Theory*, Hutchinson, London, 1980.

unite a hermeneutical conception of social experience and understanding with the desire for both an objective and an emancipatory basis for social knowledge and socio-political practice. As a part of this he has made influential criticisms of both the Germanic hermeneutical/idealist tradition and of the predominantly Anglo-American analytical/positivist tradition. His work is being seen increasingly as of fundamental importance and hence is exerting a growing influence on perceptions of the role of scientific methods and practice in social and political explanation and action. (This is despite the fact that his prose is, to say the least, dense and in need of clarification and summation.)

Habermas's objections to positivism (and hence to Popper) can be summarized by the following points. Firstly, he rejected any analytical distinction between theory, on the one hand, and its object, or experience, or history, or practice, on the other; and asserted instead the importance of a dialectical (i.e. mutually interacting) and hermeneutical unity of theory and social reality. The social life-world can be grasped only through a *circle* of understanding that *combines* 'natural hermeneutics' (i.e. commonsense unreflective understandings) with explicitly theoretical categories. The positivist approach must be replaced by an understanding and explication of existing social meanings, which rejects the dichotomy of theory and factual observation.[24]

Consequently, secondly, theory should be construed in the context of actors' prior experiences and pre-understandings of a social totality and should then be 'checked' by the theorist against his own experience in a dialectical manner. In the dialectical conception society is a non-observable whole or totality and analytical concepts and the social totality are seen as meshing together 'like cog-wheels', reinforcing each other. Social theory must be shown, and hermeneutically known, to be appropriate to this whole and not just to its empirical appearances.[25] Thus for Habermas, the

[24] J. Habermas, 'The Analytical Theory of Science and Dialectics', in Adorno, op. cit., p. 134.   [25] Ibid., p. 136.

comprehension of subjective meanings is constitutive of social theories, which should derive their categories in the first instance from the actual experience and consciousness of ordinary people and not through 'scientific' observation. Causal-analytical science tends to ignore this consciousness and, in his view, regards the appearances as being the only objective reality.[26]

However, thirdly, for dialectical thinkers such as Habermas it is insufficient to adopt either a subjectivist hermeneutical approach or a causal-analytical approach. While the meanings of the social life-world must be the starting place, those traditions must be criticized for their adequacy to the underlying level of social structural reality itself. Those subjective ideas and interpretations do depend, he believed, upon connections that people have with the *objectively real* social situation. Social understandings must not only be understood but criticized.[27]

He wished to transcend both hermeneutics and causal-analytical science by confronting the real with the *possible,* i.e. by having a *transformative* intention.[28] A practical social science must be critical of the social totality. It must both overcome the fact/value distinction and go beyond the level of cultural meanings to reveal and criticize the distorted understandings (including positivist science) which derive ultimately from the contradictions inherent in the structure of society itself. Ideological misunderstanding and the society that makes it possible are the twin targets of a critical theory.

Thus, fourthly, this meant making a critique of the dualism between facts and technical decisions that he rightly saw as lying at the heart of positivistic science. He saw this dualism as blurring the basic difference between technical and *practical* questions (a distinction traceable to Aristotle's techne/praxis dichotomy). For positivists such as Popper, the history of society, like nature, has no meaning, but an arbitrary decision can be made that it should have a certain meaning and direction

---

[26] Ibid., p. 139.   [27] Ibid., p. 139.   [28] Ibid., p. 140.

and this can then be engineered with the aid of social scientific techniques. This 'technological' attitude has the effect, Habermas believed, of eliminating questions about genuine life-practice from social science. A dialectical attitude, on the other hand, was able to draw attention to a gaping discrepancy between practical questions and merely technical tasks. It demanded the adoption instead of a conception of social theory as a potentially emancipatory critique of societies containing injustices and real contradictions. The motivation and legitimation of social science should be the solving of problems of the overall practice of life itself rather than decisions about the implementation from above of manipulative 'corrections'.[29]

The thrust of his philosophical critique of positivism was to try to undermine Popper's attempt to construct an uncontentious rock-bottom basis for objective empirical knowledge. Popper had criticized the earlier positivist attempt to found objective science upon a criterion of confirmation of hypotheses by supposedly neutral empirical observations, and had replaced this with the concept of falsifiability within a framework of decisions or conventions about what counts as a crucial test and as a basic experience. Until falsified or replaced, hypotheses and basic statements always remain provisional and no more than approximate to the truth. But, as Habermas pointed out, an agreement about basic statements can be enforced neither logically nor empirically. In the spirit of Gadamer, he argued that the scientific research process always operates in a circular manner:

One cannot apply general rules if a prior decision has not been taken concerning the facts which can be subsumed under the rules; on the other hand, these facts cannot be established as relevant cases prior to an application of those rules. The inevitable circle (cf. Gadamer) in the application of rules is evidence of the embedding of the research process in a context which itself can no longer be explicated in an analytical-empirical manner but only hermeneutically. The postulates of strict cognition naturally conceal a non-explicated pre-understanding which, in fact, they presuppose; here the detachment of methodology

---

[29] Ibid., pp. 142-3.

from the real research process and its social functions takes its revenge.[30]

Thus, he believed, the so-called basis problem would simply not arise if research were seen as embedded in a process of socially institutionalized actions through which societies sustain their life. Societies, he argued, have always had a fundamental *practical* interest in the domination of nature through societal labour, and so have achieved a consensus about the meaning of technical domination within historical and cultural boundaries. Therefore there has been an inter-subjective validity given to empirical–scientific statements due to this pre-understanding. But this prior interest becomes forgotton and the illusion of pure theory and disinterested objectivity arises.[31] Value freedom is then postulated for science and (since Max Weber) even for social science. But where technical domination may become unproblematical for natural science, for social science there are always problems since the conditions which define the situation of action cannot be divided into facts and values, ends and means. The social context is a life context and practical questions cannot be answered with a purposive rationality devoid of value judgements. Social scientific theories are not value-neutral — they are guided by pre-understandings present from the very beginning.[32]

The need to go beyond a merely meaning-comprehending hermeneutics was further demonstrated via a critique of Gadamer. Habermas developed this primarily in response to the problem of the relationship of language to the structural reality of social action and nature. Whereas for Gadamer the world is linguistically constituted through a circular process of interpretive understanding employing the prejudices or pre-understanding of enduring cultural traditions, for Habermas this was insufficient since it ignored the *objectively real* frameworks of the world:

The linguistic infrastructure of a society is part of a complex that, however symbolically mediated, is also constituted by the constraint of reality — by the constraint of outer nature that enters into procedures

---

[30] Ibid., p. 152.   [31] Ibid., p. 155.   [32] Ibid., p. 160.

for technical mastery and by the constraint of inner nature reflected in the repressive character of social power relations. These two categories of constraint are not only the object of interpretations; behind the back of language, they also affect the very grammatical rules according to which we interpret the world. *Social actions can only be comprehended in an objective framework that is constituted conjointly by language, labour and domination.* The happening of tradition appears as an absolute power only to a self-sufficient hermeneutics; in fact it is relative to systems of labour and domination. Sociology cannot, therefore, be reduced to interpretive sociology. It requires a reference system that, on the one hand, does not suppress the symbolic mediation of social action in favour of a naturalistic view of behaviour that is merely controlled by signals and excited by stimuli but that, on the other hand, also does not succumb to an idealism of linguisticity (Sprachlichkeit) and sublimate social processes entirely to cultural tradition. Such a reference system can no longer leave tradition undetermined as the all-encompassing; instead, it comprehends tradition as such and in its relation to other aspects of the complex of social life, thereby enabling us to designate the conditions outside of tradition under which trans-cendental rules of world-comprehension and of action empirically change.[33]

These latter themes of Habermas's work of the 1960s have been extensively built upon as the bases for a series of writings[34] extending up to recent years in which he has developed the concept of a critical social theory — one that is able to transcend both interpretive understanding and technical control by having an emancipatory intent. With the dominance of positivistic scientific rationality in the post-war advanced industrial world, technical control of nature and society, on an objectivist and decisionistic basis, has become the goal of theory and research. Against this he has argued that all knowledge is grounded in knowledge-constitutive interests which are represented in the cognitive/methodological

[33] J. Habermas, 'A Review of Gadamer's Truth and Method', (1970), reprinted in Fred A. Dallmayr and Thomas A. McCarthy (eds.), *Understanding and Social Enquiry*, University of Notre Dame Press, Notre Dame, 1977, p. 361.
[34] See especially the following books in English, all published by Heinemann of London in the years shown (but listed in order of German publication): *Theory and Practice* (1974), *Knowledge and Human Interests* (1972), *Toward a Rational Society* (1971), *Legitimation Crisis* (1975), *Communication and the Evolution of Society* (1979).

frameworks that prejudge the meaning of statements and establish the rules of inference. Science has a technical interest, historical-hermeneutics a practical interest, and critical theory an emancipatory interest. For Habermas, the emancipation of socialized man from the ideological and technical control of the modern state and of the international political and market systems should be the goal of a critical social theorist. And that task must begin with a demystification of the ideology that he sees as surrounding and falsely defending social and natural scientific practice as being both objective and value-free. Scientific enquiry cannot be divorced from social and cognitive interests.

This brief discussion of Gadamer, Popper, and Habermas shows, I hope, that philosophical ideas about the nature of social understanding and knowledge can have significant effects upon our conceptions of the political process and what the role of theory should be in influencing our political actions. Gadamer's hermeneutics could lead to a political passivity — a concentration simply upon understanding actors' perceptions rather than upon explaining the structural imperatives and constraints of action. A Popperian objectivist and decisionist philosophy leads to piecemeal social engineering for technical, rather than libertarian, ends. And Habermas's critical theory should lead not to utopianism (as Popper believed) but to the potential for democratic control of the structures of everyday life made possible by uncovering them from beneath their ideological disguises.[35]

IV

Echoes of the Gadamer/Popper/Habermas debate are to be found in several of the following essays since the question of the objectivity of social enquiry and hence of the foundations

[35] There has been somewhat of a convergence in the views of Popper and Habermas since the early sixties, as evidenced particularly by Popper's defence of a realism that must start with a criticism of commonsense views of the world (cf. *Objective Knowledge*, Oxford University Press, 1972, especially Ch. 2), and by their parallel development of evolutionary theories of social behaviour and human consciousness (cf. Popper, ibid., and Habermas, *Communication and the Evolution of Society*, op. cit.).

of political practice is basic to the discussion of the theory/ practice relationship. In his reflections on the relationship, which constitute the first chapter, Ralf Dahrendorf argues that the gap between them cannot easily be bridged from a theoretical point of view. The realities of political practice are such that they do not lend themselves to being organized and controlled through theory. Although theoretical reflection is important it cannot be effectively unified with practice, in his view, because of a fundamental difference in their interests and ethics. Nevertheless, they must not be forced further apart. A degree of detachment in pursuing both is necessary if they are to be combined in one person.

In discussing historical influences on the development of social theory, Tom Bottomore argues that socio-economic development and its expression in political struggles has been crucial. In other words, it was practical questions arising from new social interests that prompted systematic thought about the nature of society and politics in the first place, and the great theorists of the late nineteenth and early twentieth century Classical era of sociology, for instance, all developed their ideas within frameworks of political commitments or value orientations. In recent decades we have witnessed, in Bottomore's assessment, a retrieval of older ideas rather than a creative outburst, and in the past few years the course of political life has been downhill, prompting a narrowing and anxiety amongst theorists. But he does see light on the horizon in the form of the European protest movements.

Towards the end of his essay Bottomore draws attention to the evident philosophical turn which social science has taken in the past decade or so; and one of the foremost critics of the existing positivistic foundations of much social science and explicator of a hermeneutical alternative has been Charles Taylor, whose present essay discusses the possible roles of theory in undermining or strengthening, and in general *constituting*, the nature of our practices. (While theories do indeed have these effects, he nevertheless wishes to argue that theories can never be the simple determinants of practices.) It is these roles of theory that make the social sciences so very different, in his view, from natural science, since they are not about independent objects but rather constitute and/or transform their objects. The question then is one of the validation of theory since there cannot be a simple empirical application or test. Validation can only come, he says, from the effect upon practice — if practice is made more *clairvoyant* for the actor. But the possibility of delusion, even self-delusion, is great. There is no simple route to shared clairvoyant social understanding.

On the other hand, as Amartya Sen points out, there is a clear need for factual social knowledge — of a way of arbitrating important political and social claims. He wishes to defend the notion of 'objective social science' and does so via making important distinctions between accounts and actions, and truth and goodness. There can be objective, even truthful, *accounts* of social events and processes. But this is insufficient since we need to know if the account is a *good* one, and that depends upon what we wish a statement to be an account of. That is, he argues, we have a *use-interest* in accounts. Actions, on the other hand, are value-laden, including the actions of scientific practice and of making statements. This is the case in natural science as well as social science. Social science can arbitrate on the truth and goodness of statements but it cannot tell us why actions are performed nor provide value judgements about them.

Not only is it the case that social theory helps to constitute our practices, as Charles Taylor and Amartya Sen agree, but, as John Dunn argues, it is all we have with which to bridge the gap between our own social understandings and our experience of modern history. We may all have to be amateur social theorists, but the desire for true knowledge has produced ever stronger doses of professional causal theorizing (in a context of belief that the world may be fundamentally unintelligible) leading, in his view, to a state of ideological intoxication. Like Habermas, he believes that social theorists must learn to expect much less of explicit theory and philosophy and to take greater account of the self-conceptions held by human agents, including themselves. The gap between understanding of self and of others must be closed. Only in this way, he argues, can theory better serve practice. This conclusion is related directly by Dunn to modern British politics where the attempted imposition of official theories has led to not only great hardship but a diminution of the possibilities of social co-operation. What is needed, he believes, is a more modest and democratic vision of political authority and a closing of the gaps between amateur, professional, and official theories.

There seems little doubt that the political consensus in post-war Britain, which was remarkably strong through the 50s and 60s, has considerably weakened in recent years. As David Marquand points out, the marginalized dissenters of the 50s have now returned to centre stage. His essay analyses the elements of that consensus and adduces reasons for its breakdown. Basic to both is the level of economic performance, but, he believes, of perhaps greater importance is the regard in which government itself is held. Faith in consensual social engineering, so strong once, has been lost. No longer are economic and social science and administration (at least of the 'orthodox' sorts) seen as the means to overcome social problems. The main political parties, he believes, have reverted to their true pre-war types. However, it remains to be seen whether the spark of consensus in the population at large has died as well.

The relationship of Marxian theory to Communist practice is a perennial problem for social scientists and political practitioners. In the final essay here Wlodzimierz Brus discusses the question of the relevance of a critical Marxian theoretical outlook for comprehending and criticizing the existing communist states and thus for achieving a use-value to political practitioners within those states. He argues that a necessary condition of this is the recognition that those states contain conflicting forces and tendencies and so Marxism, a theory *par excellence* of the origins and nature of social contradictions, is then able to offer insights to reformers. Nevertheless, although Marxism is not absent from the oppositionary criticisms of the communist world it has been greatly devalued there through close association with political and theoretical failures. For Brus, this is a matter for regret since in his view Marxism can still be a valuable guide to action within communist states.

Apart from the works mentioned in the footnotes of this introduction and the following essays, readers are directed to these books which may be of use in following up aspects of the theory/politics relationship:

Hannah Arendt, *The Human Condition*, University of Chicago Press, Chicago, 1958.

T. Ball (ed.), *Political Theory and Practice*, University of Minnesota Press, Minneapolis, 1977.

Brian Fay, *Social Theory and Political Practice*, George Allen and Unwin, London, 1975.

Russell Keat, *The Politics of Social Theory*, Basil Blackwell, Oxford, 1981.

P. Laslett *et al.* (eds.), *Philosophy, Politics, and Society*, 5 volumes, Basil Blackwell, Oxford, 1956-79.

S. M. Lipset (ed.) *Politics and the Social Sciences*, Oxford University Press, New York, 1969.

Melvin Richter (ed.), *Political Theory and Political Education*, Princeton University Press, Princeton, 1980.

W. G. Runciman, *Social Science and Political Theory*, Cambridge University Press, Cambridge, 2nd ed., 1969.

Reflections on Social Theory

and Political Practice

Ralf Dahrendorf

I

The subject which Wolfson College has chosen for this year's lectures is clearly topical. But then, it has been topical for about 200 years at least, and throughout these 200 years the relationship between social theory and political practice has been characterized by the fact that it is more easily capable of engaging the passions than of producing clarity of thought and of understanding. It is a subject which lends itself to demagogy. Only this morning I read in a newspaper of a statement by a well-known American (or Canadian) economist, who said that 'Milton Friedman is as influential for British politics as Karl Marx is for Soviet politics.' Now, how influential exactly is Karl Marx for Soviet Politics? What precisely is his influence on, say, Soviet expenditure on armaments or on the invasion of Afghanistan, or on Brezhnev's statement that feeding the Soviet people is the first economic task of the country? What exactly does one mean if one says that social theorists have influenced political practice?

What I shall do in this lecture is map the subject by offering four sets of reflections and in the process raise a few sceptical questions about social theory and political practice. By political practice I shall understand the sort of thing that government ministers or perhaps Members of Parliament do, and by social theory I shall understand the sort of thing that professors do, at least certain professors — professors of political philosophy, sometimes professors of economics (after they have left the long, dark tunnel of quantitative or mathematical economics), perhaps professors of history, or even of sociology. This is all I am going to say by way of definition at this stage. Nor will I add much about the complex relationship between our attempt to understand facts (if one chooses to put it in these terms), and the other attempt to sketch possible futures, the attempt to make normative statements. I shall not deal with this classical issue in abstract terms, but it leads to the first of my four reflections.

While this may seem ever so slightly frivolous to you, let me assure you that it is meant as seriously as the other three: There are some people who seem to straddle the worlds of political practice and social theory. There are, in other words, some politician–philosophers. It can probably be said that there are (in mid-1981 at least) two members of the government of this country who might be described in these terms. One of them has directed a Research Institute, has published, has spoken about issues of social theory, and in a recent book of which he is the co-author, many of the orthodoxies of current political practice are attacked. One of the main claims of this book is that the state has become sterile. On the one hand, so it is argued, it is difficult to give reasons for the legitimacy of state action so far as, for example, the distribution or redistribution of wealth is concerned. On the other hand, the state, even if it tries to do something, is very often unable to achieve the objectives which it has set itself, and in any case state subsidies are an instrument of political action which cannot be justified in terms of political or social theory.

Now it is well known that when this author became a Minister, after a suitable though short interval, he began to act rather differently. It is well known that in the office which he holds at the moment, he listens rather more than some had expected to demands for state subsidies in particular. And he not only listens but apparently — to the surprise of some of his colleagues — he went to Cabinet meetings and asked for state subsidies to do precisely the things which he had said a year earlier in his book were useless, should not be done, and were wrong on theoretical grounds. One might argue that his humanity got in the way of his philosophy. His theory may have sounded hard; his practice increasingly seems to be what has come to be called 'wet'.

This kind of difference between theory and practice is not confined to one side of the political spectrum. On the opposition front bench there is also someone who has claims to being a social theorist. He sees himself in a long tradition of social thinkers and he likes to quote his descent from the Levellers. He likes to quote also some of the Christian roots of his own political and social thinking. And it is from this background that he states in a recent book: 'The government is the people's instrument for shaping their own destiny.' He says quite often, both in speeches and in publications, that political practice must listen to the people, that it is, as it were, from below that the truth in political practice is to be expected. Yet when he was in office and responsible for important areas of policy-making very similar to the ones of the other politician to whom I referred, there is every indication that the same man was more impressed by something which he also writes about, and that is the gigantic strides of technology towards, as he puts it, 'interdependence, complexity, and centralization'. Indeed one might be tempted to suggest that his real dream is that his practical political predilection for grand technology and his theoretical predilection for listening to the people could be joined in some mysterious way, say by having a referendum in which 85% of the people vote for having the Concorde or something of this kind.

The reason why I am giving you these examples is both simple, and rather complicated. I am not trying to criticize the two individuals I have talked about. I am trying to show by these examples (which if one went abroad could easily be multiplied) that there is, in reality at least, a strange gap between social theory and political practice. The same individuals who fully believe the things which they say and write when they engage in social theory change their posture when they occupy positions of authority. To put it differently, they suddenly find themselves subject to pressures of a kind which somehow does not have a place in the designs of social theory itself. In political practice theory somehow gets lost, — at any rate this is so for most of its tenets. There is an evident hiatus between the sort of thing one reads in books about social theory and the sort of thing one reads in the newspapers about political practice. What about this hiatus? How can we define it more closely? How has it been defined in the history of thinking about this subject?

II

This leads me to the second of my reflections (and like the subsequent ones it is about a particular person, a particular social philosopher). Probably, Hegel was the first to have given a clear, and one might say modern view of the relationship between social theory and political practice. His view is one which is still alive. In a modified form we find it in a variety of authors notably, but not only, on the European continent. If only for that reason, it is worth considering. The relevant passage in our context is of course the famous Preface to Hegel's *Philosophy of Right* in which he explains why he is trying, in this complicated and strange, yet important, book to explore the entire course of human history, or as he would put it, the march of the World Spirit through history. As he discusses his methodology, he makes one point above all, and that is that the social theorist as he thinks about the meaning of the progress of history cannot conveivably be ahead of the time at which he is writing. This claim indicates probably the fundamental meaning of Hegel's much discussed statement that 'what is reasonable is real and what is real is reasonable'. We tend to associate with the word 'reasonable' a moral, or at any rate a normative notion. Unless I misread Hegel I do not think that that is what he was trying to say. He was trying to say that the sort of thing that is thought at a particular time has a definite relationship to the sort of thing that is happening at that particular time. Theory and practice are in a one-to-one relationship, and this is so even when it is not immediately evident.

There is an interesting little discussion in this Preface of Plato's *Republic,* when Hegel argues that at first sight Plato's

Republic may look like an ideal, held before Greece or the Greek cities as an objective. In fact, however (he says) this is not so. In fact, Plato's *Republic* is no more than the extrapolation of the fundamental moral structures of the society to which Plato himself belonged. Reason cannot really transcend reality. Social theory (to put it in the language of this book) cannot really transcend political practice. Social theory reflects political practice. This is the point at which another metaphor of Hegel's becomes relevant. Hegel says that philosophy cannot actually teach the world anything. Philosophy can merely grasp the substance of he world as it is; and it can be complete, mature, and in itself perfect, only when the world itself has become mature, complete, perfect. 'The owl of Minerva only begins its flight after dusk.' There is no way for a philosopher to be ahead of the world in which he is living.

Now as I discuss political thinkers here, my interest is not only in the way in which they saw the relationship between social theory and political practice in abstract, but also in how they saw themselves. What did their view of the relationship between social theory and political practice mean for their self-interpretation, that is, for their understanding of their own position in the process of knowledge as well as the process of practice? Hegel's answer is clear. He may not have been as naive as some of his friends and enemies have thought, but there can be no doubt about the enormous arrogance of the man who not only believed that under his own eyes history had come to a final climax, but who also believed that he had the capacity to interpret this climax, to translate it into philosophical terms in such a way that after him there would be no philosophy at all, so that everybody who would follow him would be a mere epigone, a mere disciple interpreting the master.

There is not space here to go into the fascinating story of the gradual development of the Hegelian Right and the Hegelian Left. Were I to do so, I should wish to pay great attention to the extent to which Hegel had actually managed to persuade a whole generation of thinkers, at any rate in his own country, that they were mere epigones, that there was nothing new to say, that one had in fact reached the high point and the end of the march of the World Spirit through history. When Hegel says that 'the state is the reality of the moral ideal', he may not have meant the Prussian state in which he was living in every detail. Again, he may have been interpreted by some in a slightly naive way; but he certainly had it in mind that the kind of state in which he was living did mark the highest achievement of mankind and in many ways the ultimate stage of human progress.

I need not repeat at this point Popper's criticism of Hegel, but it is clear that Hegel's approach postulates a very strange relationship between social theory and political practice quite apart from the implicit dogmatism, or if you want to use the term, the 'historicism' of his analysis of world history. In a sense, social theory for Hegel is nothing but ideology in a narrow sense of the word. In a sense, social theory means that ideas merely reflect what Marx was to call the relations of production and the interests of the class maintained by them. Ideas merely reflect, in other words, reality as it is; reality with its peculiar structure of domination, and its built-in interest in continuity. Social theory cannot change things. Social theory cannot be ahead of reality nor can it be apart from it. There is no critical role for social theory in either interpretation of the term 'critical' — that is, either in the sense of, say, the Frankfurt School, or in the proper Kantian sense of the term. For Hegel, if theory deserts reality it becomes, as he puts it, vain and irrelevant.

This ideological interpretation of the relation between social theory and political practice clearly leaves out of consideration important aspects of the real role of social theory in the history of the last 200 years, and also in earlier history. Indeed, it is no accident that some of Hegel's disciples, above all those who describe themselves as the Hegelian Left, began to re-interpret this statement about the 'reasonable and the real'. They turned it round until it suited their own, more activist purposes. Their main interest was in fact for a decade or so not political theory but the interpretation of

religion. If one wanted to write a history of different interpretations of Hegel's 'what is reasonable is real, and what is
real is reasonable', one would look at Hegel's own *Life of
Jesus* as well as at the interpretations of others all the way
to David Friedrich Strauss and beyond. I shall not do that
here. The important point is that in the process of developing
Hegel something happened to the word 'is' in the statement,
'What is reasonable is real and what is real is reasonable.'
Something happened which transformed the phrase, until
in the end it became: 'What is reasonable will be real and
what is real will be reasonable.' Thus, it became one which
gave social theory a very different place in relation to the
existing social and political conditions of the time.

III

This leads me to my third reflection, and to a few words
about Marx. The obvious starting point is of course the
complicated eleventh thesis of his *Theses on Feuerbach*:
'Philosophers have merely interpreted the world in different
ways, what matters is to change it.' The statement is complicated; yet one must add that it also lends itself to the most
vicious and unfortunate interpretation. In a sense, it was so
interpreted by a non-Marxian in 1933, on that great and
horrible occasion in the history of German philosophy when
the existentialist philosopher Martin Heidegger, at the time
Rector of the University of Freiburg, made his infamous
speech in which he said that there are times in the history of
mankind when the teachers have to step back and offer the
stage to the soldiers. That infamous occasion illustrates the
simplest and also the most negative interpretation which one
might put on the eleventh thesis on Feuerbach. If one takes it
literally Marx is saying: let us forget about philosophy, let us
forget about social thought, let us concentrate on changing
things, never mind what thinkers say! Now clearly this is not
the whole truth of Marx's thesis. Clearly the young philosopher
who was thinking about Feuerbach was not trying to say that
philosophy is useless. Moreover, he was not even trying to
abolish the connection between philosophy and reality, the

idea of which he had inherited from Hegel. He was really trying to say that as long as economic, social, and political conditions are in some fundamental sense wrong, philosophy is bound to be wrong as well. It is only once social and political conditions have been put right that philosophy can be right as well.

If one talks about right and wrong, if one uses moral categories in connection with Marx, one is in a sense always doing him an injustice; Marx was not thinking in moral terms. He thought in terms of the historical inevitability of processes which would take place whether anyone liked them or not. Yet this is only three-quarters of the truth, because Marx himself engaged in political practice. To be sure, if one looks at his vain attempts at political action closely and biographically, one soon finds that in his case, the relationship between social theory and political practice was not much more effective than in the two cases which I cited at the beginning of this lecture. That fact apart, the eleventh thesis on Feuerbach is in fact saying: there is a connection between social theory and political practice, but the connection is not as Hegel saw it. It cannot be held that social theory today is right and true because political reality is right and true. Political reality is, on the contrary, such that it makes social theory today wrong, intrinsically wrong; the conditions of capitalist society do not allow a social theory which is right — with one exception.

Once again we come to the other side of the picture: how does a man who has this kind of approach interpret his own position? How does he see his own social theory? Many of you will be familiar with the places in *The Communist Manifesto*, and one or two other publications of Marx's before 1848, where he argues that there are certain periods of change, certain times of turmoil, in which a small number of individuals are, as it were, liberated from the social conditions which dominate the lives of most. By being thus liberated (Marx argues), they are capable of insights which are not available to a majority, and were not available to anyone in earlier periods of historical development. In other words, Marx has a rather daring, not to say unconvincing, personal theory which enables him to say that while all that matters at the moment is to change things, there are some, notably he and those who believe in his theories, who have an idea of where the process of historical development is going to lead. They are therefore capable of engaging in social theory rather than having to wait for political practice.

I cannot claim to be an expert on the details of the Marxist tradition, but it would seem that Marx's personal way out of a Hegelian position turned upside down, was the beginning of one particular Marxist tradition which is still very much with us today. This tradition likes to emphasize the importance of theory and claims there is a sense in which one should be able to say that theory and practice are not two separate activities but are intertwined in what is called a dialectical relationship. Indeed, some say that they are united to the point of becoming one and the same thing. Theory as the recognition of the historical process is practice, and such practice cannot be without theory. It is only a step from this school of thought to Gramsci who saw an even more pronounced role for theory in the historical process.

With Marx and Marxism, we are once again faced with an ambitious attempt to find a solution to the strange problem of the hiatus between theory and practice. What Marx was trying to say is that theory is a part of a process which is as yet unfinished, and not merely the reflection of a completed process. But if one thinks this approach through, and looks at the reality of attempts to link theory and practice in this particular way, there is little left in it that could be called convincing. However far authors have gone in pursuing the Marxian approach, they have not been able to take the step from social theory to political practice. Their very terminology remains somewhat unreal. It is all very well to talk about 'communication without constraint', but enunciating the words does not make the concepts real. The wall between thought or theory and action or practice remains as high and as strong as ever. On closer inspection, the statements that theory and practice are one, or are tied together in a dialectical union, are either imprecise, or mystical truths for the initiated.

IV

Before I try to pull the various strands of this argument together, let me turn to my fourth point, and to yet another German author, Max Weber. Let me say just a few words about his two important speeches of 1919 which are relevant in this connection. The first, on 'Science as a Vocation', contains the famous argument that politics does not belong in the lecture hall and that one should distinguish very clearly between what the scholar does and what the politician does. Why should one? Now this lecture of 1919 is in one respect rather amusing, for the fairly modern, reasonably progressive, democrat Max Weber actually produced as one of his reasons the rather Willhelminian argument that in the lecture hall where the teacher faces his listeners they have to keep quiet, and the teacher has to speak. This does not have to be so, of course. One wonders whether Weber would have changed his position had he experienced the type of teaching which is more characteristic of universities in this country, where there is a lively interchange in classes between teachers and students. However that may be, the deeper reason given by Weber for his distinction between scholarship (and one might well substitute 'social theory' for this) and what he calls politics is of course the famous statement that no amount of scientific inquiry can prove a value judgement, that however hard we try to research or argue a case in the framework of scientific inquiry, this will not enable us to give sufficient reasons for the truth of a value judgement. For that reason Weber wanted to keep science entirely on one side and politics on the other.

In the second great lecture of 1919 on 'Politics as a Vocation', Max Weber makes a distinction between an 'ethic of conviction' which refers to absolute values and is not prepared to accept any compromises in reality, and an 'ethic of responsibility', that is, a moral approach which judges particular situations in a pragmatic fashion, not leaving absolute moral standards totally out of consideration but at the same time not letting them govern one's political actions. He says that politics must of necessity be governed by an ethic of responsibility, because if it were governed by an ethic of conviction it would in fact be unpolitical. Politics, he says, is carried on with one's head, but most certainly not with one's head alone. Politics is a passionate and at the same time a pragmatic exercise. It is not in any sense the application of social theory. The two are in important respects separate. Thus, for Max Weber, contrary to Hegel and to Marx, the hiatus between social theory and political practice seems to be almost absolute.

One hesitates to try to advance an intellectual explanation for the personal breakdown which Max Weber suffered before the First World War; but it may not be entirely wrong to say that he insisted, at the risk of his sanity, on a distinction which it was exceedingly difficult to maintain for a man whose passions were as much scientific as they were political. After all, Weber himself not only gave these lectures on 'Science as a Vocation' and 'Politics as a Vocation', but also wrote his big work *Economy and Society*, which by his own standards is entirely value-free and scientific; and at the same time he published highly polemical topical articles such as those in which he attacked in a clearly political fashion unconditional submarine warfare in the First World War. There were thus two souls in the man, and yet he kept on insisting that they must not in any way be confounded. They must remain totally separate. I have often wondered whether such a position is sustainable.

To give you an illustration of why I think it is not, let me quote one of the great social theorists who is fortunately still amongst us: Karl Popper. I remember him telling me soon after his 75th birthday how delighted he was that he had received messages of congratulation from both the German Chancellor Schmidt, and the Leader of the Opposition Kohl; from the Austrian Chancellor Kreisky, and the then Leader of the Opposition Taus. I said to him, 'Professor Popper, does it not worry you that the same theories seem to attract opposition and government in very similar ways? Do you not believe that your theories should above all be accepted and used and believed in by those whose political predilections you share?' But he said, 'no, not at all. I could not care less. I know what I believe and make it clear, and that is good enough.'

Is there anything wrong with this position? In a sense, no-one could disagree with Popper: of course people know what he stands for. But on a deeper level, a more difficult question arises. It has been the subject of fierce discussions in connection with Weber, though not with Popper. It is this: if one distinguishes in this extreme way between the world of social theory and the world of political practice, there is a great danger that by implication at least, one leaves the world of political practice to unargued, unreasoned, and possibly unreasonable decisions. The question of whether Weber was a hidden decisionist or whether he was actually prepared to support a constitution which would enable a charismatic individual to do whatever he wanted to do, will undoubtedly remain unresolved. It is clearly unfair to suggest that Weber, a liberal and democrat, was in any sense, however indirectly, responsible for what happened thirteen years after his death. But there is the important and quite fundamental question of what political practice would be like, if people did not insist on demanding reasons for what was being done, if they did not insist on argument, in other words, on a degree of social theory in the execution as well as the formulation of policy.

In a recent German book the splendid political philosopher who is now teaching in Zürich, Herman Lübbe, argues convincingly that it would be wrong to charge Max Weber with pure decisionism, because one must distinguish clearly between reasons that are given for the substance of political decisions (and in his view Weber never objected to that), and reasons that are given for the practical or factual validity of political decisions. Insofar as their validity is concerned, Lübbe argues that it is perfectly legitimate to take a pragmatic or institutional view, and not to insist upon permanent discussion and permanent argument. This is a possible interpretation of Weber. Whether it solves the problem of the hiatus to which I have alluded time and again, is another matter. One can of course try to turn the hiatus between social theory and political practice itself into a theory, as Weber has done; but the price is high. It is a human, an existential price; people are torn apart by the two ethics, and even by Lübbe's distinction between theory and pragmatic action. This is not to suggest that there is an easy solution. If there is any one conclusion from this discussion of authors (who will certainly be discussed further in this book) it is that the gap between social theory and political practice exists. Indeed as I turn to a few concluding remarks, you will find that they fall far short of offering a bridge across the gap, a solution of the problem.

V

The first of my concluding remarks is that so far as one can see, there is no theoretical bridge across the gap between social theory and political practice which is fully satisfactory. The unity of theory and practice is in many ways an empty phrase. It could mean that theory is a mere mirror of practical politics, a mere derivation of real social and economic conditions, which in my view sells theory short because there should be the possibility of detachment, of anticipation, of developing images of the future which are not simply a reflection of the present. From this point of view, it is a degradation of theory to talk about the unity of theory and practice. And if one looks at the phrase from another end and says that all practice has to be theoretical somehow, and both consider authors who claim this and their effect on reality, the conclusion is inescapable that they make practice ineffective. They ignore the realities of political practice, the pressures which exist, the institutions which constrain action, the needs of power. The ethic of responsibility has its own claim on human action, and no amount of conviction can make it disappear. In other words, there is no theoretical bridge between theory and practice, and we should not delude ourselves that there is such a bridge.

On the other hand — and this is the second concluding remark which I want to make — it seems to me that Weber's position cannot really be sustained. It is wrong to go out of one's way to try to widen the gap. It cannot do any harm to political practice to be forced to argue, or to social theory to take into account the experiences of political practice. There is a need to expose the two to each other. I do not want to ride a personal hobby horse in this connection, but will just mention that the kind of institution which for example exists in the form of the Brookings Institution, where those who are engaged in social theory and in political practice meet, cannot but be a good thing for both, What is more, I think it is both conceivable and desirable that in our own complicated societies we have an increasing number of people who are prepared to straddle the two worlds, who are not worried about straddling them, people who have tried to understand some of the intrinsic problems of either world, and then try somehow, although not at the same moment, to unite social theory and political practice in their own lives.

This leads me to a final and third point. You may remember that as I began I said that there are probably at least two ministers in the present Cabinet whom one might say combine a degree of social theory with their political practice, and those who do remember will also remember that so far I have only referred to one. The other one is, I think, a political practitioner who enjoys thought and occasionally embarks on flights of theory, though he does not do so with much pretence. The result of his very personal combination of theory and practice is a rather pleasing irony. He may not be the most effective politician, he may not be the greatest social theorist, but his irony is, in itself I think, a value in a world where irony is not abundant. In my view, a social theorist soaked in practice is quite likely to reach the same conclusion; a degree of detachment, a degree of ironical reflection on those complicated relations between those who think, and those who do things.

Social Theory and Politics in the History of Social Theory

Tom Bottomore

I

Lytton Strachey, in one of his portraits of historians, remarked that three qualities make a historian, '. . . a capacity for
absorbing facts, a capacity for stating them, and a point of
view'; and he continued: 'The latter two are connected,
but not necessarily inseparable. The late Professor Samuel
Gardiner, for instance, could absorb facts, and he could state
them; but he had no point of view; and the result is that his
book on the most exciting period of English history resembles
nothing so much as a very large heap of sawdust.'[1]

It is apparent to everyone that the landscape of the social
sciences too is abundantly furnished with heaps of sawdust,
large and small. But this is not all that it contains. Those
thinkers who mapped the territory, invented the concepts,
sketched or elaborated the hypotheses and theories, of the
modern social sciences — from Adam Smith to Keynes and
Schumpeter, from Saint-Simon to Durkheim and Max Weber,
from Bentham to Gramsci — had a very definite 'point of
view', and one which was, for most of them, pre-eminently
political.

I have said 'most of them' because I do not intend to claim
that the point of view of every important social theorist
without exception has been directly and centrally political.
Two examples will show that this is not the case. The work
of Simmel, which undoubtedly contributed ideas of very great

---

[1] Lytton Strachey, *Portraits in Miniature and Other Essays,* Chatto and Windus,
London, 1931, pp. 169-70.

significance to sociological theory, was certainly not oriented
in any narrow sense to political issues, although one part of
it, devoted to a critique of modern culture — and notably the
Philosophy of Money — does bear particularly upon the
problems of capitalism, has a clear relation to Marx's theory,
and had, for instance, a substantial influence upon the later
Marxist writings of Lukács. The structural anthropology of
Lévi-Strauss may seem to be even more distant from political
concerns, and determined by purely theoretical interests,
although here too there is an explicit connection with Marxism,
and even a declared intention to contribute, through the
study of myth, to a theory of superstructure only hinted at
by Marx.[2]

With these reservations, however, I do want to assert that
the development of modern social theory as a whole — that
is to say, of the theoretical social sciences — has been closely
and inextricably linked with socio-economic development
and its expression in political struggles. The very origins of
these sciences — and in particular, of the attempts to construct
a general social science, sociology — are inseparable from the
political crises in Western Europe in the eighteenth century
resulting from the rapid growth of a capitalist economy and
the emergence of new social interests. This is nowhere more
clearly revealed than in the *Encyclopaedia* of Diderot and
d'Alembert, which was intended to be not simply a summation
of modern knowledge, but a particular advocacy of social
science, and a contribution to the advance of the democratic
movement.[3] Nor is a sociologist likely to disregard, in the
light of Robert Nisbet's study of the formative period of his
discipline, how profoundly some of its basic themes and
concepts were affected, in an opposite sense, by the conservative reaction to the French Revolution.[4]

Elsewhere I have examined in a broader context the way in

[2] C. Lévi-Strauss, *The Savage Mind*, Weidenfeld and Nicolson, London, 1966,
p. 130.
[3] See the account given by René Hubert, *Les sciences sociales dans l'Encyclo-
pédie*, Félix Alcan, Paris, 1923.
[4] Robert A. Nisbet, *The Sociological Tradition*, Basic Books, New York, 1966.

which, during the nineteenth century, social theory became ever more intimately linked with political doctrines and with social movements which aimed to bring about major changes in the organization of society, whether by the constitution of new nations or the renewal of old ones, by the overthrow of capitalism or the achievement of wide-ranging reforms to mitigate its social consequences. On one side, therefore, social theorists became pre-eminently concerned with what they conceived as the vital political problems of their age; and on the other side, social theories themselves came to be seen, in quite a new way, as a necessary foundation of political doctrines, and as providing elements which could be directly incorporated into the programmes of social movements and parties. The latter tendency went so far indeed that some Marxist thinkers, beginning with Engels, came to speak of 'scientific socialism'; that Karl Mannheim could later discuss the 'prospects for a scientific politics' in terms of a possible synthesis of world views accomplished by the intellectuals; and that still more recently, from an opposite standpoint, the 'critical theorists', especially Jürgen Habermas, have been led to concentrate one main arm of their criticism upon the very process by which, as they see it, political issues have been transformed into scientific and technical problems. Yet we should note that even in this last case the new style of politics which is advocated again depends heavily upon an underlying social theory.

Such considerations, however, lead too far afield. On the present occasion I shall confine myself strictly to the historical connections between social theory and politics, disregarding the broader philosophical questions which arise about the relation between theory and practice. Even within these limits I shall not attempt to provide a comprehensive historical panorama of such connections, but merely illustrate the theme by considering the theoretical and political orientations of some important sociological thinkers (that being the intellectual history which I know best), and then conclude with some brief reflections on the recent history of politics and social thought.

II

The most conspicuous example of a very close connection between social theory and politics is to be found, of course, in the life and work of Karl Marx. In the first place, Marx's basic conceptions, as they took shape in the 1840s, brought together in a single scheme of thought a great part of the politically oriented social theory of the preceding half century: the humanist philosophical anthropology of Feuerbach; the ideas of the French socialists, notably the Saint-Simonians; and the analyses, by political economists, of the early stages of industrial capitalism. Second, in its developed form, Marx's theory (notwithstanding its partial or complete exclusion, over long periods of time, from the official academic world) came to occupy a pre-eminent place in social thought as a whole, and to have a more profound and dramatic direct influence upon the course of political events than any other single body of ideas has ever done. In its social and political aspects at least, the century which has passed since Marx's death deserves to be called the 'century of Marxism'.

In the course of that century, however, the intense debate within and about Marxism has produced numerous reinterpretations of the theory, in response to intellectual criticism and the impact of events; and as one major element in these ever-renewed assessments, there have been quite diverse accounts of the manner in which Marx's theory itself was originally constituted. My own interpretation of this historical genesis is that the moment of illumination came, the discovery and working-out of the central idea occurred, between 1842 and 1844 and is recorded in Marx's writings of that period: in his critique of Hegel's theory of the state and philosophy of right, in his essay on the Jewish question, and in the *Economic and Philosophical Manuscripts*. The discovery was the idea of the proletariat as the most important social and political factor in modern society. This discovery then led Marx to an analysis of the situation of the proletariat in terms of property, production, and exchange, and to the eventual assertion of class struggle as the principal dynamic element in social life. These writings represent the first crucial achievement, though of course still incomplete, of Marx's intellectual project formulated in the letter he wrote to his father in November 1837 in which, describing the new direction that his life was taking, he says that he was 'greatly disturbed by the conflict between what is and what ought to be' and that eventually 'starting out from idealism . . . I hit upon seeking the Idea in the real itself'. The proletariat was this idea in the real world.

But let us be clear what this means. I do not think it can be maintained that Marx, in a single, unitary cognitive act, grasped both the idea and the real, value and fact; for the idea — the moral idea of human emancipation — had already taken definite shape in Marx's thought several years before he discoverd the proletariat or began his analysis of the structure of capitalist society, and indeed already guided his support for the democratic radical movement while he was editor of the *Rheinische Zeitung* in 1842. Rather we should say that Marx, from a political standpoint already largely formed, saw in the proletariat a social group which, because of its real situation in society, was bound to engage in a struggle for emancipation, or as he wrote in the 'Critique of Hegel's Philosophy of Right': 'a sphere of society which cannot emancipate itself without emancipating all the other spheres of society'. It seems quite accurate to say, therefore, as did Hendrik de Man in his review of the *Economic and Philosophical Manuscripts* when they were first published in 1932, that these manuscripts show 'more clearly than any other work the ethical–humanist themes which lie behind Marx's socialist convictions . . . and behind his whole scientific work'.[5]

This view of Marx's *political* approach to a conception of the central importance of the proletariat, in theory and in practice, is confirmed, I think, by the fact that it was quite possible to identify and define the proletariat in a very similar way, and yet arrive at an entirely different theory of the development of modern society. This was the case notably with Lorenz von Stein, whose book on the social movement

---

[5] In *Der Kampf*, XXV, 1932, pp. 224-9, 267-77.

in France, published in 1842,[6] provided the first substantial
sociological interpretation of the proletariat, and very probably had a considerable influence in crystallizing Marx's
own thought on the subject, but who went on to espouse the
cause of a reformed capitalism in which the state would play
an important role in improving the condition of labour (and
in fact to outline in his later writings a conception of the
welfare state). Marx, on the contrary, proceeded from *his*
conception of the proletariat to construct, in one direction,
a theory of history as the history of class struggles, and in
the other, a theory of the structure, and the contradictions,
of modern capitalist society; the fruits of these endeavours
being first expounded in a more or less systematic way in
The German Ideology (1845) and *The Poverty of Philosophy*
(1847).

It has been generally recognized that the concepts of class
and class struggle occupy a central position in Marx's social
theory, and I have argued that his conception of the revolutionary proletariat was in fact the starting-point for the
construction of his theoretical system as a whole. Equally,
however, it became a principal object of critical examination
and questioning, for the validity of the theory came to be
seen as depending crucially upon the truth of its empirical
statements about the historical development and political
role of the proletariat in capitalist societies. There are, to
be sure, many different levels at which the theory can be
criticized, and many aspects of it which have been subjected
to particular criticism and revision — its ordering of the stages
of historical development; its account of the relationship
between the state, civil society, and the economy; its theory
of ideology; its analysis of capitalist production — but its
practical political theses about the modern class struggle and
the transition to socialism have been, throughout the period

[6] Lorenz von Stein, *Geschichte der sozialen Bewegung in Frankreich von 1789
bis auf unsere Tage* (1842. 3rd revised and enlarged edition 1850.) English translation of major sections of the book, edited with an introduction by Kaethe
Mengelberg, *The History of the Social Movement in France, 1789–1850,* Bed-
minster Press, Totowa, N. J., 1964.

since Marx's death, a point of conspicuous difficulty, both
for Marxists of various persuasions from Bernstein to Marcuse,
and for critics of Marxism from Durkheim and Weber to
Schumpeter.

I think it is evident that this difficulty has increased over
time, for the course of world politics in the twentieth century
has not conformed at all closely with Marx's analysis. Two
streams of events have run counter to his expectations.
One is the occurrence, in 1917, of a proletarian revolution
without a proletariat, and the consequent necessary sub-
stitution of a party for a class; or perhaps on a longer
view the first manifestation, in a specific historical situation, of a quite different kind of struggle, that of the
intellectuals on the road to class power, as Konrád and
Szelényi have described it in a recent study of East European
societies.[7]

The other phenomenon, which from a Marxist standpoint constitutes rather a stream of non-events, is the absence
in the developed capitalist countries of a revolutionary proletariat, or even of any significant trend in that direction
except on very rare occasions. Hence Hilferding's posing
of the problem in his last unfinished manuscript, when
he reflected upon four decades of political action, that
'. . . nowhere has *socialist* consciousness taken hold of the
entire working class';[8] C. Wright Mills's dismissal, especially
in the American context, of the Marxist conception of the
proletariat as being an untenable 'labour metaphysic';[9] and
Marcuse's pessimistic conclusion in *One-Dimensional Man*
that there are no longer demonstrable real forces in established

[7] George Konrád and Ivan Szelényi, *The Intellectuals on the Road to Class Power*, Harvester Press, Brighton, 1979.

[8] Rudolf Hilferding, *Das historische Problem* (1941). Unfinished manuscript, first published, with an introduction by Benedikt Kautsky, in *Zeitschrift für Politik* (new series) I, 4, December 1954, pp. 293–324. English translation of part in Tom Bottomore (ed.), *Modern Interpretations of Marx*, Basil Blackwell, Oxford, 1981, pp. 125–37.

[9] C. Wright Mills, 'The New Left', in *Power, Politics and People*, Oxford University Press, New York, 1963, p. 256.

society through which a critical theory could translate its rationality into historical practice.[10]

The immense importance of the actual course of events for Marxist social theory can be brought out more sharply by carrying out the kind of imaginative reconstruction which Max Weber proposed in order to establish the significance of particular historical events and to arrive at a synthesis of the real causal complex, by disregarding (as he said) one or more of those elements of 'reality' which are actually present and by the mental construction of a course of events which is altered through a modification of one or more conditions.[11] Let us suppose, then, that at some time during the first two decades of this century a revolutionary proletariat had actually emerged in at least some of the major capitalist countries of Western Europe, that a socialist consciousness had taken hold of even a majority of the Western working class. Such a proletariat might have prevented the outbreak of the First World War, or brought it speedily to an end; the Central European revolutions at least might have succeeded; and socialist societies might have been established in forms which corresponded more closely with Marx's own conception of a future society of associated producers who would regulate their interchange with nature rationally and humanely. In that case his social theory would have received a powerful confirmation,[12] and however much debate might have taken place about particular aspects of his historical and structural analyses, or about questions of method, what has emerged as its central weakness — the point around which all the critical and self-critical discussions of Marxism revolve, more closely or more distantly — would not have appeared; namely, the disjunction between the idea and the real.

[10] Herbert Marcuse, *One-Dimensional Man*, Routledge and Kegan Paul, London, 1964, esp. pp. 254–7.

[11] Max Weber, 'Critical studies in the logic of the cultural sciences', in Edward A. Shils and Henry A. Finch (eds.), *The Methodology of the Social Sciences*, The Free Press, New York, 1949, pp. 171–5.

[12] Even Lukács's 'revolutionary, utopian messianism' (as he later called it in his preface to the 1967 edition of *History and Class Consciousness*), embodied in his view of Marxist theory as the expression in thought of the revolutionary process itself, would then have had some justification.

To avoid any misunderstanding let me make clear the scope of the preceding argument. It does not deny the existence in Western capitalist societies of a class system and of class struggles. It does not raise at all the question of the validity of Marx's analysis of the main tendencies of capitalist production. In that sphere I think Marx was broadly right, and is increasingly recognized to have been right, in his general argument to the effect that in an unregulated capitalist economy there are powerful forces tending to produce recurrent crises, and ultimately a more profound 'collapse' or long-term stagnation. (Keynes's analysis in the *General Theory* came to similar conclusions.[13]) Marx's analysis is not invalidated by the fact that in the post-war period Western capitalism has not so far experienced serious crises or prolonged stagnation, for this can be accounted for precisely by massive state intervention and the widespread adoption of Keynesian policies to counteract its inherent tendencies. From the other side we can perhaps find a confirmation of Marx's analysis in the present state of the British economy, which seems to show in a particularly striking way that free market capitalism of a pre-Keynesian kind is simply not viable. More generally, however, there doubtless remains a serious question about whether even 'organized capitalism', as Hilferding called it, can in the long run avoid stagnation.

But this, as I have said, lies outside the range of my present argument which is concerned not with the economic fate of capitalism, but with its political destiny, and more particularly with Marx's expectation that it would be superseded, as a result of a conscious political struggle, by a new and 'higher' form of society. It is the failure of that political struggle to reach the intensity that Marx anticipated, or some slackening of the struggle, or a substantial change in its character, which has led to what many have called a 'crisis in Marxism' and to the search in more recent social theory for new forces of change; that is to say, for a new embodiment of the idea in the real.

[13] See the comparison between Keynes and Marx in Karl Kühne, *Economics and Marxism*, Macmillan, London, 1979, Vol. II, Part IV, pp. 237-60.

III

Before pursuing that theme, however, let us consider how the relation of theory to politics is treated in some other classical schemes of sociological thought which, together with Marxism, still provide today the main theoretical framework for a general science of society. If Marx aimed to bring social science and politics into the closest possible association, Max Weber, it might be said, set out to distinguish and separate them as sharply as possible. That surely is the meaning, in particular, of his two essays on science as a vocation and politics as a vocation.[14] Indeed it is; but only in a sense which needs to be carefully specified. Neither in principle nor in his scholarly practice did Weber espouse the idea of a 'value-free' science to be set against a 'value-laden' politics, or construe the distinction between science and politics as an impermeable barrier.

It would be wholly inconsistent with Weber's underlying conception of the social world to suppose that any kind of human activity could be 'value-free'. On the contrary, it was a fundamental tenet of his thought that all human action consists precisely in endowing the world with meaning and value. The distinction between science and politics, therefore, is one between two different value spheres; science being the sphere of cognitive values (or what Weber also refers to as rationalization and intellectualization), politics the sphere of those values related to struggles for power and the exercise of power (hence also the use of violent means).

Thus, in 'Science as a vocation', Weber refers, in his colourful language, to the 'gods of the various orders and values': 'We live today [he says] as did the ancients when their world was not yet disenchanted of its gods and demons, only we live in a different sense. As Hellenic man at times sacrificed to Aphrodite and at other times to Apollo, and above all, as everybody sacrificed to the gods of his city, so do we still today. . . . Fate, and certainly not "science", holds sway

---

[14] In H. H. Gerth and C. Wright Mills (eds.), *From Max Weber*, Kegan Paul, Trench, Trubner, London, 1947, pp. 77-128, 129-56.

over these gods and their struggles'. In 'Politics as a vocation'
he observes that these different 'gods' may come into conflict
with each other; the god (or demon) of politics, he says,
'. . . lives in an inner tension with the god of love', and '. . . this
tension can at any time lead to irreconcilable conflict'. Equally,
he could have claimed that there is an inner tension between
the god of politics and the god of science.

More important in the present context, however, is the
fact that Weber did not regard the different value spheres as
being entirely closed off from one another. On the contrary,
in his essay on the meaning of 'value freedom' he sets out
plainly his view that it is 'cultural (evaluative) interests
[which] give purely empirical scientific work its direction';[15]
and again in his essay on 'objectivity' that 'there is no
absolutely "objective" analysis of culture or . . . "social
phenomena", independent of special and "one-sided" points
of view according to which . . . they are selected, analyzed
and organized for expository purposes'.[16] And he concludes
this essay with a strong reaffirmation of the influence of
extraneous cultural values upon science: 'All research in the
cultural sciences in an age of specialization . . . will regard
the analysis of [its] subject matter as an end in itself . . . It
will cease to be aware of its rootedness in ultimate value ideas.
And it is good that this is so. But a time comes when the
atmosphere changes. The significance of the unreflectively
used points of view becomes uncertain . . . The light of the
great cultural problems has moved on. Then science too
prepares to change its standpoint and its conceptual apparatus
. . . It follows those stars which alone are able to give meaning
and direction to its labours'.[17]

It is worth noting here, although I cannot pursue the
question at length, that in these methodological writings
Weber emphasizes very strongly the influence of other cultural
values upon science, and does not give equal consideration to
the reciprocal influence of science upon other values; as, for

---

[15] Shils and Finch, op. cit., p. 22.   [16] Ibid., p. 72.
[17] Ibid., p. 112. Translation modified.

example, the English sociologist L. T. Hobhouse did in
formulating his idea of a rational ethic, or as later positivists
and near positivists were inclined to do. The general impression
produced is that Weber assigns the value realm of science to
a subordinate place in some unstated hierarchy of values; and
this is connected, I think, with his intention to undermine
the position of the economic interpretation of history (as it
was expounded in the orthodox Marxism of the German
Social Democrative party) by arguing that this is only one
possible interpretation among others, equally valid, made
from different extraneous value standpoints. By contrast, in
his substantive studies of modern capitalism, Weber lays great
stress upon the general cultural influence of science as a
major element in the process of rationalization, intellectualization, and disenchantment of the world; and in the essay on
science as a vocation he makes one brief reference to this
phenomenon as the 'fate of our times', while declaring at the
same time that he himself 'hates intellectualism as the worst
of devils'.

At all events, it is evident that Weber's own studies were
guided by a well-defined point of view, a distinctive value
orientation, which on occasion he expressed in forthright
terms. Karl Löwith, in his monograph on Weber and Marx,
argues that there underlies the work of both thinkers a
'philosophical anthropology' (that is, a non-empirical conception of man in modern capitalist society) which centres,
in Marx's case, upon the idea of 'alienation', in Weber's case,
upon the idea of 'rationalization'.[18] Löwith's exposition,
although it succeeds admirably in elucidating some undoubtedly important features of their respective value orientations,
seems to me to remain at too abstract a level, and not to
bring clearly into view the contrast between Marx and Weber
in the selection they made of a reference point for their
theoretical analyses in the world of real political struggles.
For Marx, as I have indicated, this reference point was the

[18] Karl Löwith, *Max Weber and Karl Marx*, English trans., Allen and Unwin, London, 1982.

proletariat; for Weber, as has been made increasingly clear by recent studies, it was the nation state.

In his inaugural lecture at Freiburg, at the outset of his scholarly career, Weber himself stated bluntly that the basic principle of his political theory was 'the absolute primacy of the interests of the nation state', which constitute an 'ultimate standard of value' in both politics and economics; and he went on to describe the general context of social theory as one in which 'we should recognize that the unification of Germany was a youthful folly . . . which it would have been better not to undertake at all, in view of its cost, if this was to be the conclusion, and not the starting point, of Germany's striving to become a world power.'[19] Weber maintained this point of view throughout his life, reiterating it explicitly in two of his last writings – his essay on 'Parliament and Government in a Reconstructed Germany'[20] and his lecture on socialism[21] – and he analyzed a great range of problems concerning the economy, bureaucracy, democracy, and socialism, in relation to the paramount need for strong national leadership. As Mommsen has said, he 'never envisaged any other world than his own, characterized by the rivalry of nation states'.[22] This overriding commitment to the interests of the German state did not, however, prevent Weber from expressing, in his Freiburg lecture and on many later occasions, very vigorous criticism of the political weaknesses of the German bourgeoisie, and of the leadership of various bourgeois parties. In just the same way, a Marxist social theorist, from the standpoint of the 'absolute primacy' of the interests of the proletariat, might well take an extremely critical view of the political capacities of the working class and of working class political parties; and this would certainly

[19] In *Gesammelte Politische Schriften,* 3rd enlarged edn., Tübingen: J. C. B. Mohr, Tübingen, 1971, pp. 1-25.

[20] English trans. in *Economy and Society,* Bedminister Press, New York, 1968, Vol. III.

[21] English trans. in W. G. Runciman (ed.), *Max Weber: Selections in Translation,* Cambridge University Press, Cambridge, 1979.

[22] Wolfgang Mommsen, *The Age of Bureaucracy: Perspectives on the Political Sociology of Max Weber,* Basil Blackwell, Oxford, 1974.

be preferable — especially in the light of the changed conditions
of the late twentieth century — to the fabrication of a
mystique of the proletariat which has characterized much
Marxist thought.

IV

In a different, less immediately obvious, and less extreme
way Durkheim's sociology was also oriented to the interests
of a nation state. At first sight it may appear that we have to
do here with a particularly single-minded and determined
attempt to create a 'value-free' science, a strictly 'objective'
and 'scientific' sociology. But this is not the case. From the
beginning Durkheim had two aims, one theoretical, the
other practical. His practical aim was to contribute to the
'regeneration of France', and to accomplish this aim, as he
said in the conclusion of his series of lectures on socialism,[23]
by discovering 'through science the moral restraint which can
regulate economic life'. That is why, in *The Rules of Sociological Method,* published at this time, he devoted a long
chapter to the distinction between the 'normal' and the
'pathological' in social life, in an attempt to establish a
scientific basis for value judgements; and to set out more
explicitly and systematically a distinction which he had
already employed in discussing the 'abnormal forms of the
division of labour'.[24]

It would be more accurate to say, therefore, that Durkheim's primary aim, throughout his life, was to discover and
implement a new moral regulation of society (in social and
political conditions which I shall discuss in a moment), and
that the construction of a social science was subordinated to
this end; although this social science was also, of course,
from his point of view, essential to it. Certainly, Durkheim's
concern with morality seems to have become increasingly
prominent in his work; and Marcel Mauss, in his memorial

[23] English trans. *Socialism*, Collier Books, New York, 1962.
[24] English trans. *The Rules of Sociological Method*, University of Chicago
Press, Chicago, 1938, Chapter III.

essay published in the first issue of the new series of the *Année Sociologique* (1925),[25] records that in the last year of his life, in spite of his illness, Durkheim began to write the book on morality which was (in Mauss's words) 'the purpose of his existence and the foundation of his thought'.

But Durkheim's idea of moral regulation had a political sense which it is important to bring to light. In his work as a whole, unlike that of Marx and Weber, there is relatively little direct analysis of political power. Only in a course of lectures given in the 1890s and first published long after his death[26] did he formulate systematically, if briefly, his conception of the state as 'a special organ responsible for elaborating certain representations which are valid for the collectivity. These representations are distinguished from other collective representations by their higher degree of consciousness and reflection. . . . Strictly speaking, the state is the very organ of social thought'. In these lectures some substance is given to the idea, which otherwise appears in a rather shadowy form in Durkheim's discussions of social solidarity, of the state as the supreme moral regulator of society, and we begin to apprehend more clearly the political meaning of his theory.

For Durkheim, as later for Talcott Parsons, the state embodies and expresses a national collective will, and has primacy over all other interests and purposes in society. His nationalism emerges most distinctly in some of his writings during the First World War, which reflect (as do Weber's on the other side) the struggle between France and Germany for supremacy in Europe as a basis for world power;[27] but it is

[25] Reprinted in Marcel Mauss, *Oeuvres*, Éditions de Minuit, Paris, 1969, Vol. III, pp. 473-99.

[26] *Leçons de sociologie: physique des moeurs et du droit* (1950). English trans. *Professional Ethics and Civic Morals*, Routledge and Kegan Paul, London, 1957.

[27] See especially, *'Germany Above All': German Mentality and the War*, Armand Colin, Paris, 1915. For an account of how Tönnies and Sombart (as well as Weber), on the German side, were also overcome by patriotic fervour, see Arthur Mitzman, *Sociology and Estrangement: Three Sociologists of Imperial Germany*, Alfred A. Knopf, New York, 1973, pp. 129-31, 261-4.

expressed indirectly in many other contexts: in his view of
class struggle as a temporary (and of course undesirable)
'abnormality', in his rejection of the idea of working class
internationalism, and in his emphasis upon the need for a
reformed educational system which would provide adequately
for the moral education of the young generation to equip
them for their tasks in the collective life of the nation. It was
in the sphere of education that Durkheim's ideas probably
had their greatest practical impact; so that a Marxist critic
could say in the 1930s that '. . . in the name of Durkheim's
science, school children are taught to glorify *la patrie francaise,*
to justify class collaboration, to accept everything, to commune in the cult of the flag . . .'.[28]

Thus Durkheim's political commitment was very similar to
that of Max Weber; but it developed in different circumstances
and so took a different form. Whereas for Weber the political
problem consisted in maintaining or reviving the impetus given
to German expansion by the unification of Germany under
Bismarck, for Durkheim the problem was to recreate a powerful and stable French nation state in the conditions resulting
from defeat in the war with Prussia, from the Paris Commune
and the violent class antagonisms that it bequeathed, and
from the continuing unresolved conflict, carrying back to the
Revolution, between the monarchist and catholic right and
the republican anti-clerical left — in short, to provide a secure
foundation for the Third Republic. In their specific contexts,
therefore, and in the more general context of the power
struggles among European states, both Weber and Durkheim
were guided in the elaboration of their social theories by their
commitment to a strong nation state, based upon capitalist
production (though this was to be in some sense a reformed
capitalism in which welfare provisions would allay class
antagonisms), and modern in the sense that it would be a
democratic state (though this conception was qualified in
Weber's case by the notion of the 'plebiscitarian leader', and
was not fully analyzed by Durkheim at all). There remain, of

[28] Paul Nizan, *Les chiens de garde*, Rieder, Paris, 1932, p. 97.

course, some important differences between Durkheim and Weber. Durkheim did not assert, as Weber did, that the interests of the nation state constitute an absolute standard of value. On the contrary, in his brief analysis of the state, he referred explicitly to values common to humanity as a whole which were 'higher' than those of any particular nation, although he placed the actual embodiment of such values in social institutions in a very distant and uncertain future, and so concluded that for the time being such values could only be expressed, in fact, by nation states. Undoubtedly, though, Durkheim was a more moderate nationalist than was Weber; and we might say, in a more general comparison of their political commitments, that Durkheim was more liberal and more democratic than Weber, but also less of a realist, in the sense that his conception of the state as a moral agency largely excluded recognition of its character as an instrument of power and violence.

V

Social theory, like other cultural phenomena, follows an undulating course, with peaks of creativity and troughs of stagnation; and as I have portrayed it here — dealing only with the most general forms of social thought — its movement seems to conform quite closely with that of political events. The most important theories were provoked, and were given their distinctive character, by the two most significant political elements in the development of European societies in the nineteenth century; first, in the earlier part of the century, the formation of an industrial working class and its irruption into political life, and second, in the latter part of the century, the rapid growth of capitalism, organized in increasingly powerful nation states, with the larger states engaged in fierce struggles to extend their economic territory and to enhance their position as world powers.

Of course, this is a very broad sketch, to which a great deal of qualifying detail needs to be added. The social sciences have not all developed in exactly the same way, and I have been primarily concerned here with sociological thought.

Even in a particular field it cannot be claimed that every major theoretical conception has been affected to the same extent by the political milieu, and I pointed to some exceptions at the outset in referring to the work of Simmel and Lévi-Strauss. More generally, the account which I have given of Marx, Weber, and Durkheim should not be understood to mean that their social theories are nothing more than disguised political ideologies. Like Schumpeter[29] I take the view that a history of social theories is more than just a history of ideologies; and I do so for much the same reasons, deriving in part at least from Weber's distinction between 'value spheres'. What becomes a central problem for any social theorist is certainly conditioned on one side by the historical development of cultural phenomena, and especially political struggles, and on the other side by partly — but not exclusively — individual value orientations which not only affect his concentration upon this or that conceptualized aspect of social life but colour in some way his whole thought. But the activity of theoretical construction is also guided by the values of science (that is, by cognitive values) which, even if they are not the only values that are effective, even within science, are always powerful and often predominant. Hence, whatever the starting-point and coloration of social thought, a social theory can still claim — and the claim has to be evaluated within science — to capture and represent in thought some real constellation of facts or events in the social world. Hence, as I have suggested, however we may now judge the political orientation of Marx's thought to the class struggle of the proletariat and his anticipation of its outcome, it can still be claimed that his analysis of the historical tendency of capitalist production is broadly correct, and at the same time such a claim can be rationally and empirically disputed.

The broad historical picture that I have sketched might, of course, be questioned more fundamentally from a point of view quite different from my own. In fact, some other

[29] J. A. Schumpeter, *History of Economic Analysis*, Allen and Unwin, London, 1972, Part I, Chapter 4.

attempts to provide a general historical account of modern social thought seem to me to show considerable similarities with my own version. A good example is furnished by H. Stuart Hughes's study of the 'reorientation of European social thought between 1890 and 1930',[30] which surveys one of the important periods that I have discussed. Notwithstanding considerable differences in the approach to the subject, and in much of the detail of interpretation, there is in the first two chapters, which set the stage for his discussion, a clear focus of attention upon the central questions with which I have been concerned here; namely, the consequences of the rapid development of industrial capitalism in the European nation states, and the extent of class conflict in those states. Thus, the first chapter on the revolt against positivism in the 1890s takes up the theme — very familiar again today — of hostility to industrialism, science, and technology, and to an increasingly organized, centralized, and to some extent militarized, form of social life; a theme which acquired great prominence in German sociological thought, notably in Tönnies' criticism of modern capitalism (much influenced by Marx) in terms of a contrast between 'community' and 'society', and of course in Weber's interpretation — considered by many scholars to be the *leitmotiv* of his whole work — of the development of modern society as a process of 'rationalization'.

The second chapter of Hughes's book, on the critique of Marxism, mentions — although this aspect is not much emphasized — some of the conditions which provoked the numerous critical attacks on the Marxist theory, the most important probably being the rapid growth of European socialist parties inspired by Marxism as a political force which could no longer be ignored, and which signalled for many people the imminent demise of the existing society. But it also notes, in an opposite sense, the first indications of a questioning of Marx's idea of the revolutionary role of the

[30] H. Stuart Hughes, *Consciousness and Society: The Reorientation of European Social Thought 1890–1930*, Alfred A. Knopf, New York, 1958.

working class, in the controversies engendered by Bernstein's writings and the 'revisionist' movement.[31] At this time, Hughes argues, there began a process of disentangling the two elements — scientific investigation and practical precept — in Marx's thought, exemplified in a particularly interesting and idiosyncratic form in the work of Georges Sorel. Such attempts at disentanglement, and counter-attempts to restore the unity of theory and practice, have continued of course up to our own day, and they have always been vitally affected by the course of political struggles.

May we conclude then that there is a close and continuous relationship between social theory and politics which is plainly demonstrated in the historical development of the social sciences? Raymond Aron, in a post-war appendix to his book on German sociology,[32] suggested that the revival of sociology in Germany after the Second World War illustrated the 'cruel observation that historical misfortunes encourage the development of the social sciences'. This characterization, although it is quite close to my own view in some respects, seems rather too narrow. What incites fresh thought in the social sciences, I would say, is not so much particular misfortunes, but a more general crisis in which an existing form of society and political regime has to confront an array of new problems and challenges arising from new social interests and aspirations, and in which there is at the same time a positive development of new conceptions of social life, flowering on occasion into the idea of creating a new type of society or civilization.

These negative and positive elements were clearly apparent during the most recent period of great animation in the social sciences, the decade of the 1960s. For the revival of social theory at that time was most obviously connected with a

[31] For other accounts of this movement see Peter Gay, *The Dilemma of Democratic Socialism: Edward Bernstein's Challenge to Marx*, Columbia University Press, New York, 1952; George Lichtheim, *Marxism: An Historical and Critical Study*, Routledge and Kegan Paul, London, 1961, Part V, Chapter 6; Leszek Kolakowski, *Main Currents of Marxism*, Clarendon Press, Oxford, 1978, Volume II, Chapter IV.
[32] Raymond Aron, *German Sociology*, Heinemann, London, 1957, p. 135.

series of crises — the crisis of deStalinization in Eastern Europe, and the later stages of the crisis of decolonization in the Western countries (to mention only the most prominent features) — but equally with a mood of optimism, amounting to a conviction among some social groups that a real advance was once again possible, that human emancipation and participatory democracy were actually within our grasp. But while this was indeed a very animated period I do not think, in retrospect, that it was outstandingly creative. What happened in social theory was not so much an elaboration of new ideas as a preliminary retrieval of old ones, to be seen in the general revival of Marxist thought, and more particularly in the re-emergence of the 'critical theory' of the Frankfurt School. Perhaps the main factor in this situation was that the social movements which stimulated new theoretical reflection were so short-lived, in some cases brutally suppressed. Looking back it may seem that the political ideas of these movements were as ephemeral as mini-skirts or the music of the Beatles. In the following decade, at all events, the course of political life was downhill all the way; from the Soviet invasion of Czechoslovakia, through the establishment of divers tyrannies such as in Chile, to a general state of disillusionment and reaction in a large part of the world. This has gone so far indeed that we now almost take it for granted that any major political event can only be some new act of barbarism, pointing towards the ultimate barbarity of nuclear war.

Such conditions are hardly propitious for the emergence of new and imaginative social theories. Social research comes to be guided by the short-term practical needs of policy making within the framework of the dominant, largely uncriticized, political doctrines; while social theory tends to become anxiously preoccupied with the solidity of its own foundations and the validity of its traditions, that is to say, with problems in the philosophy and history of science.

But there is no compelling reason to suppose that the present political context of social theory will itself endure. On the contrary, there is a good deal of evidence to suggest that in Europe at least the influence of the radical movements of the 1960s has persisted, and is now growing again, notably in the anti-nuclear and ecology movements. If this revival continues it will engender more intense political debate as well as new courses of action, and these in turn will provide the context and substance for a renewal of social theory.

Political Theory and Practice

Charles Taylor

I

We could understand a great deal about our political thought and practice if we could get clear the distinction between political theory and the theories of natural science. And that means clear both about what they are like as theories, and about how they relate to political action.

Natural science gives us a model which is tolerably clear. Theory, say physical theory, gives us a picture of the underlying mechanisms or processes which explain the causal properties and powers of the things we are familiar with. In many cases, this picture of the underlying reality turns out to be surprising, or strange or paradoxical, in the light of our ordinary commonsense pre-understanding of things. We have to revise our views radically about the nature of the things which surround us in order to explain what goes on.

But part of what is involved in having a better theory is being able more effectively to cope with the world. We are able to intervene successfully to effect our purposes in a way that we were not able before. Just as our commonsense pre-understanding was in part a knowing how to cope with the things around us; so the explanatory theory which partly replaces and extends it must give us some of what we need to cope better. Theory relates to practice in an obvious way. We apply our knowledge of the underlying mechanisms in order to manipulate more effectively the features of our environment.

There is a constant temptation to take natural science theory as a model for political theory: that is, to see theory as offering an account of underlying processes and mechanisms of political society, and as providing the basis of a more effective planning of political life. But for all the superficial analogies, political theory can never really occupy this role. It is part of a significantly different activity. So much might be widely agreed. Just what activity, and how it differs, is harder to say.

The superficial analogy is evident enough. The basic question of political theory could be put this way: what is going on? What is really happening in society? We need to ask this question because the ordinary commonsense understanding of what is going on is inadequate. And the answers offered frequently are surprising, strange, even shocking for ordinary understanding.

Of course, this definition of the question might be challenged. Don't we ask and answer a host of questions in political theory? And isn't there a distinction between two different departments of this theory, 'normative' and 'explanatory'? But we can agree with the immense diversity of political theory, and even agree provisionally with the normative/explanatory distinction, and still recognize that the fundamental issue concerns what is really happening in political society. I hope that this claim will appear more convincing as I proceed.

But let us look first at the disanalogy with natural science, that is in the nature of the commonsense understanding that theory challenges, replaces, or extends. There is always a pre-theoretical understanding of what is going on among the members of society, which is formulated in the descriptions of self and other which are involved in the institutions and practices of that society. A society is among other things a set of institutions and practices, and these cannot exist and be carried on without certain self-understandings.

Take the practice of deciding things by majority vote. It carries with it certain standards, of valid and invalid voting, and valid and invalid results, without which it would not be the practice that it is. For instance, it is understood that each participant makes an independent decision. If one can dictate to the others how to vote, we all understand that *this* practice is not being properly carried out. The point of it is to concatenate a social decision out of individual decisions. So only certain kinds of interaction are legitimate. This norm of individual independence is, one might say, constitutive of the practice.

But then those who carry on this practice must, in general and for the most part, be aware of this norm and of its application to their own action. As they vote, they will generally be capable of describing what they are doing in something like these terms: for example, 'we are voting properly', or 'something dubious is going on there', or 'that's foul play'. These descriptions may be mistaken, of course, but that is another matter. The point is that self-awareness in this dimension is an essential condition for a population's engaging in this practice. If none of those involved had any sense of how their behaviour checked out on this dimension, then they would not be involved in *voting*. Perhaps they would simply be playing some game that they barely understood, which involved marking papers.

In this way we can say that the practices which make up a society require certain self-descriptions on the part of the participants. These self-descriptions can be called constitutive. And the understanding formulated in these can be called pre-theoretical, not in the sense that it is necessarily uninfluenced by theory, but in that it does not rely on theory. There may be no systematic formulation of the norms, and the conception of man and society which underlies them. The understanding is implicit in our ability to apply the appropriate descriptions to particular situations and actions.

In a sense, we could say that social theory arises when we try to formulate explicitly what we are doing, to describe the activity which is central to a practice, and to articulate the norms which are essential to it. We could imagine a society where people decided things by majority vote, and had a lively sense of what was fair and foul, but had not yet worked out explicitly the norm of individual dependence and its rationale in the context of the practice. In one clear sense, their doing so would amount to a step into theory.

II

In fact the framing of theory rarely consists simply of making some ongoing practice explicit. The stronger motive for making and adopting theories is the sense that our implicit understanding is in some way crucially inadequate or even wrong. Theories do not just make explicit our constitutive self-understandings, but extend, or criticize, or even challenge them. It is in this sense that theory makes a claim to tell us what is really going on, to show us the real, hitherto unidentified course of events.

We can distinguish some of the forms this kind of claim can take: it may be that we see what is really going on only when we situate what we are doing in a causal matrix which we had not seen or understood. We are all familiar with a classic example of this kind of claim: the proletarian is engaged in making contracts with independent owners of capital to exchange his labour power for wages. What he fails to see is that the process in which he so engages by contract is building the entrepreneur as owner of capital, and entrenching his own status as an agent without other recourse than selling his labour for subsistence. What looks like an activity between independent agents is actually part of a process which attributes to these agents their relative positions and status. So says Marxist theory.

In this case the constitutive self-understanding which is upset is that which belongs to the activity of making and fulfilling contracts between independent agents. On one level, this self-understanding is not wrong; and it is certainly constitutive of a capitalist society in Marx's view. That is, workers have to understand themselves as free labourers in order to be proletarians. But when we see it in the broader matrix, its significance is in an important way reversed. What seemed a set of independent actions are now seen as determined and forced. What seemed like one's making the best of a bad job is now seen as a yoke imposed on one.

But the Marxist theory also upsets the political self-understanding described above, that of decision by majority vote in 'bourgeois' society. For in fact the matrix of the capitalist economy severely restricts the choices open. Options which reduce profitability threaten everyone with economic decline, and potential mass unemployment. These severe limits will in general mean that the very options which are offered to voters will be pre-shrunk as it were, to be compatible with the continued unhampered operation of the capitalist economy. So once again, what looks like a collective decision freely compounded out of autonomous individual choices is in fact structurally determined. Or so the story goes according to this theory.

There is thus one kind of claim; altering or overturning our ordinary implicit understanding, on the grounds that our action takes place in an unperceived causal context, which confers on it a quite different significance. But there are also theories which challenge ordinary self-understanding and claim that our actions have a significance we' do not recognize, not by virtue of an unperceived causal context, but because we are allegedly blind to certain dimensions of significance themselves.

Plato's picture of the decay of the polis in the *Republic* provides a well-known example of this: what seems like the competition of equals for place and fame is in fact a fatal abandonment of moral order, engendering a chaos which cannot but deepen until it must be brought to an end in tyranny. The inner connection between democracy and tyranny is hidden from the participating citizen, because he cannot understand his action against the background of the true order of things. He just stumbles from one to the other.

In our day, there are a number of theories abroad of this kind. We can think, on one hand, of Freudian-influenced theories, which portray the real motivations of political actors and the real sources of political power and prestige quite differently from the rational, instrumental, unilitarian forms of justification that we usually provide for our choices and allegiances. Or think, on the other hand, of the picture often presented by opponents of the culture of growth: we blind ourselves to the importance to us of a harmony with nature and community in order the more effectively to sacrifice these to economic progress. Indeed, some of the most influential of these theories critical of growth find their roots in Plato. We have only to think of the late E. F. Schumacher.

Critical theories of this kind often propound some conception of false consciousness. That is, they see the blindness in question as not just ignorance, but in some sense motivated, even wilful. This is not to say that theories which portray our action as taking place in a broader causal context cannot also invoke false consciousness. Marxism is a case in point. They must do so to the extent that the causal context is one that ought normally to be evident, so that its non-perception is something we have to explain. But this need for a special explanation of non-perception becomes the more obvious when what we allegedly fail to appreciate is the moral or human significance of our action.

There is a special class of this kind of theory of unperceived significance which I want to single out here, because it will be important for my later discussion. These are theories of what we could call common meanings. They hold a view common to the effect that the nature of social life will depend crucially upon the kind of significances which are shared among members. By 'shared significance' here I mean something different and stronger than mere parallel or convergent significance. I do not mean cases where members perceive the significance of their actions and social processes in the same way. I am talking of cases where the significance is something that they appreciate together, and where this common appreciation is itself constitutive of the significance.

The well-known example is the one central to the tradition of civic humanism, the citizen republic. This takes its character from its laws; so that the citizen's action takes on a crucial significance by its relation to the laws: whether it tends to preserve them, or undermine them, to defend them from external attack, or to weaken them before enemies, and so on. But this significance is essentially a shared one. The laws are significant not *qua* mine, but *qua* ours; what gives them their importance is not that they are a rule I have adopted. The culture in which this could confer importance is a quite different one, a culture of individual responsibility, perhaps even incompatible with that of the republic. Rather the laws are important because they are ours. And this cannot simply mean, of course, that our private rules converge on them; their being ours is a matter of our recognizing them as such together, in public space. In other words, that the significance is shared is a crucial part of what is significant here. Public space is a crucial category for republics, as Rousseau saw.

Some theorists in our tradition have taken shared significance seriously. They include, I believe, Aristotle, Machiavelli, Montesquieu, Rousseau, Tocqueville; in our day, Arendt and Habermas, to mention just two. A rather diverse lot. But a central notion they share is that having important meanings in common puts us on a different footing with each other, and allows us to operate as a society in a radically different way. The thinkers of the civic humanist tradition were interested in how men could become capable of acting together in a spontaneously self-disciplining way, the secret of the strength of republics. Machiavelli, indeed, saw this as the secret of strength in the most direct and crude military terms. But the general insight shared by all thinkers of this cast is that our way of acting together is qualitatively different when we act out of shared significance. This is the basis of what Hannah Arendt called 'power', attempting to redefine the term in the process.

This can be the basis for a certain class of theories of unperceived significance, to the effect that the way our actions either strengthen or undermine our common meanings escapes us. So that we can, for instance, be in the process of destroying our republican political community blindly. The destructive significance of our action is lost on us. Of course, this kind of theory can appear paradoxical, since it seems to be supposing that some significances which are shared are not fully perceived. But I hope to show later on how this paradox can be dissolved, and that a theory of this kind must be taken quite seriously.

In any case, we have seen several ways in which theory can claim to tell us what is really going on in society, challenging and upsetting our normal self-descriptions, either through identifying an unperceived causal context of our action, or by showing that it has a significance that we fail to appreciate. And I suppose, in order to make this list a trifle less incomplete, I should add that theories are not necessarily as challenging to our self-understanding as the ones I have mentioned here. They can have the function just of clarifying or codifying the significance which is already implicit in our self-descriptions, as I indicated earlier. For instance, some elaborate theory of the order of being, and the related hierarchy of social functions, may fit perfectly into the practices of a stratified society. It may simply codify, or give explicit expression, to the habits of precedence and deference already in being.

And theories of the causal context can play the same unchallenging role. Since the eighteenth century, our culture has been saturated with theories of the economy, which show the train of transactions effecting the production and distribution of goods as following laws. These purport to make us aware of regularities in the social process of which we would otherwise be ignorant. But this knowledge may just complement our self-understanding, not overthrow it. Not all theories of political economy are revolutionary. This was Marx's complaint.

Relative to the 'democratic' picture of ourselves above as deciding matters through majority vote, certain theories of the economy are not at all upsetting. They present us, for instance, with a picture of 'consumer sovereignty', matching in parallel our political image of voter sovereignty. These theories of the economy promise to show us how to design policies which are more effective, which intervene with greater awareness and hence success in the underlying processes of the economy. To do this, as with any application of technology, we have to respect the scientific laws governing this domain. But this is not seen as making a sham of choice, as in the Marxist picture.

These theories all claim, challenging or not, to tell us what is really going on. What kind of a claim is this? They do not deny, at least never entirely, that what we see ourselves doing in our surface understanding is also going on. So what is special about the processes theory reveals? Presumably that this description captures the really significant process. But significant for what?

Following the analogy with natural science, we are naturally tempted to reply: significant for explanation. Theory identifies the important explanatory factors, the ones which can account for stability or change, revolution or decay. But it is plain that some of these theories are also claiming to identify what is humanly or morally significant. Are these two quite different kinds of things, the proper objects of two quite distinct forms of theory?

The belief that this is so underlies the distinction between normative and explanatory theory mentioned above. I think that any attempt to constitute these two as independent departments of theory is bound to fail. One can convince oneself of this by coming to see that claims both about explanatory and human significance have consequences for our view of the nature of human motivation. Any theory of either kind involves identifying important goals, purposes, desires, and aspirations of human beings. And this means that only certain explanatory views are compatible with certain normative views, those, namely, which are grounded in similar notions of human nature. I have tried to argue this else-where[1] and will not try again here. I would just like to be able to take as granted the unity of explanatory and normative theory though what I say below will serve as part-argument.

For the interweaving of normative and explanatory will emerge, I think, from another angle, if we look at the dis-analogy I mentioned above between natural science and politics, viz., that the commonsense view which theory upsets or extends plays a crucial, constitutive role in our practices.

---

[1] 'Neutrality in Political Science', in P. Laslett and G. Runciman (eds.), *Philosophy, Politics and Society*, 3rd Series, Basil Blackwell, Oxford, 1966.

This will frequently mean that the alteration in our understanding which theory brings about can alter these practices; so that, unlike with natural science, the theory is not about an independent object, but one that is partly constituted by self-understanding.

Thus a challenging theory can quite undermine a practice, by showing that its essential distinctions are bogus, or have a quite different meaning. What in the 'democratic' picture looks like unconstrained choice is presented by Marxist theory as unyielding domination. But that means that one of the constitutive norms of the practice of majority decision is shown as in principle unfulfillable. The practice is shown to be a sham, a charade. It cannot remain unaffected. People will treat this practice and the connected institutions (such as legislatures) very differently if they become convinced of the challenging theory. But this is not a matter of some psychological effect of further information. The disruptive consequences of the theory flow from the nature of the practice, in that one of its constitutive props has been knocked away. This is because the practice requires certain descriptions to make sense, and it is these that the theory undermines.

Theory can also have the radically opposite effect. An interpretation of our predicament can give added point to our practices, or show them to be even more significant than we had thought. This is, for example, the effect of a theory of the chain of being in an hierarchical society. Relative to our 'democratic' picture, some theory which showed that important economic or other issues are up for grabs, and await our determination, would have the same heightening effect.

III

A theory can do more than undermine or strengthen practices. It can shape or alter our way of carrying them out by offering an interpretation of the constitutive norms. Let us start again from our picture of 'democratic' decision by majority rule, the picture which is implicit in our practices of elections and voting. There are a number of ways of understanding this process. We can see this by contrasting two of them.

On one hand, we have an atomist model, which sees society as a locus of collaboration and rivalry between independent agents with their individual goals. Different social arrangements and different dispositions of society's resources affect the plans of members differently. So there is naturally struggle and competition over policy and position. 'Democratic' decision-making allows people equal input and weight in determining how things are disposed, or tolerably near to this. This view might be made more sophisticated, so that we see the political system as open to 'inputs', in the form of 'demands' and 'supports', and as producing as output an 'authoritative allocation of values', in which case we could develop quite a complex intellectual grid to describe/explain the political process, in the way that David Easton has done.[2]

Quite different from that would be a republican model, issuing from one of the theories of common meanings mentioned above. From this standpoint the atomist theory is ignoring one of the most crucial dimensions of social life, viz., the degree to which the society constitutes a political community, that is, the kind and degree of common meanings. A society in which all goals were really those of individuals, as they are portrayed in the atomist scheme, would be an extreme case, and a degenerate one. It would be a society so fragmented that it was capable of very little common action, and was constantly on the point of stasis or stalemate.

A society strong in its capacity for common action would be one with important shared significances. But to the extent that this was so, the process of common decision would have to be understood differently. It could not just be a matter of how and whose individual demands are fed through to the process of decision, but would also have to be understood at least partly as the process of formulating a common understanding of what was required by the shared goals and values. These are, of course, the two models of decision that are

[2] David Easton, *The Political System*, New York, 1953; and *A Systems Analysis of Political Life*, New York, 1965.

invoked in the first two books of the *Social Contract*. Rousseau's aim is to show how one can move from the first to the second; so that we no longer ask ourselves severally what is in our individual interest (our particular will), but rather what is the proper content of the general will. The proper mode of social choice is where the policy selected is agreed upon under the right intentional description; what is crucial is that it be agreed upon as the right form of a common purpose, and not as the point of convergence of individual aims. The latter gives us merely *la volonté de tous*, whereas a true community is ruled by its *volonté générale*.

Rousseau thus presents in very schematic sketch the notion of a certain form of social decision, which for all those thinkers who fall into the civic humanist tradition is seen as normative. Societies fail to have true unity, cohesion, strength to the extent that their decisions emerge from the will of all as against the general will. The immense gap between the atomist and general will theories is thus clear. What the second sees as a defining feature of the degenerate case is understood by the first as a structural feature of all societies. Which is just another way of saying that the crucial dimension of variation among societies for the second is quite unrecognized by the first.

But it ought to be clear that the general acceptance of either of these models will have an important effect on the practices of social decision. These practices may be established in certain institutions, which may be the same from society to society, or in the same society over time. But within this similarity, the way of operating these institutions will obviously be very different according to whether one or the other model is dominant, i.e., has become the accepted interpretation. Where the atomist model is dominant, decision-making of the general-will form will be severely hampered, suppressed, and confused. Where on the contrary some self-understanding of common meanings is dominant, the scope for will-of-all decisions will be circumscribed within the bounds of explicit common goals.

Indeed, there might be no quarrel with this point about

the effect of these theories. The problem might be seeing why their effect is not greater; why, for instance, the dominance of atomist theories does not put paid to general-will decisions althogether. The answer lies in the fact that a theory is the making explicit of a society's life, i.e. of a set of institutions and practices. It may shape these practices, but it does not replace them. So even though some feature may find no place in the reigning theory, it may still be a constitutive part of a living practice.

The notion of the general will can be seen as a way of formulating the constitutive norm of decision-making for communities with shared meanings. Even if this norm remains unformulated and unrecognized, it may still be that the community retains certain shared meanings. These will still be central to certain of its practices, for example, to the kinds of arguments that are acceptable/unacceptable in public debate, even if they are unformulated. Of course, they will be very much damped down, and much less vigorous in public life than where they are publicly acknowledged. And they will certainly be in danger of eclipse. But they may nevertheless be still operative. Theory can never be the simple determinant of practice. I want later to claim that something like this gap between theory and practice is true of our contemporary society.

This is the striking disanalogy between natural science and political theories. The latter can undermine, strengthen, or shape the practice that they bear upon. And that is because (a) they are theories about practices, which (b) are partly constituted by certain self-understandings. To the extent that (c) theories transform this self-understanding, they undercut, bolster, or transform the constitutive features of practices. We could put this another way by saying that political theories are not about independent objects in the way that theories are in natural science. There the relation of knowledge to practice is that one applies what one knows about causal powers to particular cases, but the truths about such causal powers that one banks on are thought to remain unchanged. That is the point of saying that theory here is about an independent object. In politics on the other hand, accepting a theory can itself transform what that theory bears upon.

Put a third way, we can say that while natural science theory also transforms practice, the practice it transforms is not what the theory is about. It is in this sense external to the theory. We think of it as an 'application' of the theory. But in politics, the practice is the object of theory. Theory in this domain transforms its own object.

This raises difficult problems about validation in political theory. We cannot think of this according to a simple correspondence model, where a theory is true to the extent that it correctly characterizes an independent object. But it is also totally wrong to abandon the notion of validation altogether, as though in this area thinking makes it so. The fact that theory can transform its object does not make it the case that just anything goes, as we shall see below. Rather we have to understand how certain kinds of changes wrought by theory are validating, and others show it to be mistaken.

IV

Before trying to show how this is so, I have to acknowledge that a powerful current in our culture resists strongly the idea of political theory as transforming its object. Partly because of the very puzzlement about validation just mentioned, and partly for other reasons, the temptation has been strong to assimilate political theory to the natural science model. This theory would then aspire like physics to yield knowledge about the unchanging conditions and regularities of political life. This knowledge could be applied to effect our ends more fully should we find occasion and justification.

Of course, it is difficult to present in this light theories which claim to identify the true significance of our actions. And so the attempt is usually made with theories of the causal context. The various theories of political economy have tended to be of this form: certain consequences attend our actions regardless of the intentions with which they are carried out. So no alteration in our self-understanding will alter *these* regularities. Our only way of changing the course of things is by *using* these regularities to our own ends. In short, practice must apply the truths of theory. We have here exactly the relation of natural science.

We have been brought generally to consider economics as a science of this kind. Monetarism is true or false as a proposition about how certain economic transactions concatenate with others. These effects accrue to certain types of intervention regardless of the intentions, awareness, and state of mind of those who are responsible for them. Greater awareness can alter the result only if it can yield a different way of making these laws work for our purposes. So when it all does not work according to plan, one calls a master technician back from America.

Perhaps there is some justification for this as far as economics is concerned. There are certain regularities which attend our economic behaviour, and which change only very slowly. But it would be absurd to make this the model for social theory in general and political theory in particular.

First, there are cultural conditions of our behaving according to these regularities. Economics can hope to predict and sometimes control behaviour to the extent that it can be confident that in some departments of their lives people will behave according to rather tightly calculable considerations of instrumental rationality. But it took a whole vast development of civilization before the culture developed in which people do so behave, in which it became a cultural possibility to act like this; and in which the discipline involved in so acting became widespread enough for this behaviour to be generalized. And it took the development of a host of institutions, money, banks, international markets, etc., before behaviour of this form could assume the scale it has. Economics can aspire to the status of a science, and sometimes appear to approach it, because there has developed a culture in which a certain form of rationality is a (if not the) dominant value. And even now, it fails often because this rationality cannot be a precise enough guide. What is the rational response to galloping inflation? Economics is uncertain where we ordinary agents are perplexed.

Second, we could not hope to have a theory of this kind, so resistant to our self-understanding (relatively resistant, as we have seen) outside the economic sphere. The regularities are there, and resistant, to the extent that behaviour responds to narrow, circumscribed considerations. Economic behaviour can be predictable as some game behaviour can be; closely circumscribed in a given domain. But for that very reason, a theory of this kind could never help to explain our motivated action in general.

Various attempts to explain political behaviour with an economic-model theory always end up either laughable, or begging the major questions, or both. They beg the question to the extent that they reconstruct political behaviour according to some narrowly defined conception of rationality. But in doing this, they achieve not accuracy of description of political behaviour in general, but rather they offer one way of conceiving what it is to act politically, and therefore one way of shaping this action. Rather than being theories of how things always operate, they actually end up strengthening one way of acting over others. For instance, in the light of our distinction above between atomist and general-will constructions of democratic decision-making, they help to entrench the atomist party. Setting out with the ambition of being natural-science-type theories of an independent reality, they actually end up functioning as transforming theories, as political theories normally do, but unconsciously and *malgré elles*. They thus beg the interesting question: is this the right transforming theory? because they cannot even raise it; they do not see that it has to be raised.

If on the other hand, they try to avoid this partisanship by becoming rather vague and general in their application, allowing just about any behaviour to count somehow as rational, then they become laughable. Theories of this kind generally hover between these two extremes. An excellent example is the conversion theory of politics mentioned above in connection with the name of David Easton.

Another type of theory which was allegedly natural-science-like, and resistant to self-understanding, was the clutch of structuralisms which recently were in vogue in Paris. The confusions which these were prey to are now too well documented to need rehearsing here. Theories of this kind get whatever plausibility they have by equivocating over the notion of structure. On one understanding, that drawn from linguistics which is one of the important sources of this kind of view, a structure is constantly renewed and changed in action, and hence is not resistant to changes in self-understanding. This type of structure is then confused with the unchanging resistant type, whose model is the laws of natural science, or else (in the case of Althusser) Marx's theory of political economy. The uncreative confusion between these two notions of structure has already cost the republic of letters enough time and treasure, so that the best policy now is perhaps one of benign neglect.

What emerges from this is that the model of theory as of an independent object, or as bearing on an object resistant to our self-understanding, has at best only partial application in the sciences of man. It can apply only in certain rather specialized domains, where behaviour is rather rigid, either because largely controlled by physiological factors, or because a culture has developed in which what is done in a given department is controlled by a narrow range of considerations, as in games or (to some degree) economic life. But this could never be the general model for social science, and certainly not that for a science of politics.

V

This brings us back to the question of validation. What is it for a theory to be right? We cannot just reply that it is right when it corresponds to the facts it is about. Because, to oversimplify slightly, political theories are about our practices (as well as the institutions and relations in which these practices are carried on), and their rise and adoption can alter these practices. They are not about a domain of fact independent of, or resistant to, the development of theory.

If theory is about practices here, then what makes a theory right is that it brings the practice out in the clear. And what this leads to is that the practice can be more effective in a certain way. Not just in any way, but in the way practices can be when we overcome to some degree the muddle, confusion, cross-purposes which affect them as long as they are ill-understood. To have a good theory in this domain is to understand better what we are doing; and this means that our action can be somewhat freer of the stumbling, self-defeating character which previously afflicted it; our action becomes less haphazard and contradictory, less prone to produce what we did not want at all.

In short because theories which are about practices alter those practices the proof of the validity of a theory comes in the changed quality of practice that it enables. Let me introduce terms of art for this shift of quality, and say that good theory enables practice to become less stumbling and more clairvoyant.

We should note that attaining clairvoyant practice is not the same thing as being more successful in our practices. It may be that there is something deeply muddled and contradictory in our original activity, as for instance Marxism would claim about the practices of 'bourgeois' democracy. In which case, theoretical clarity is not going to enable us better to determine our own fate within the context of bourgeois institutions. Rather what the theory will have revealed is that this enterprise is vain; it is vitiated at the very base. But practice can be more clairvoyant here because we can abandon this self-defeating enterprise, and turn to another goal which makes sense, i.e. revolution. Of course, if we bring this off, we shall have been more successful overall; but not in the practices we originally set out to understand, which we have on the contrary abandoned. And just getting the right theory does not ensure that we can bring off the revolutionary change. We may just be stymied. Still, if the theory is right we would be capable of more clairvoyant practice, which in this case would just consist of our abandoning the muddled, self-stultifying effort to determine our fate freely within the structures of the capitalist economy.

This makes validation a very difficult and controversial matter when it comes to political theory. How do we determine that practice has become more clairvoyant? We usually have a distinct sense that it has when we make the transition from more muddled to more clairvoyant. But two big questions obviously obtrude: how do we know we are not deluding ourselves? And how do we convince others who have not made the transition?

Of course, these questions could arise about any theories, even those of natural science. But they cause particular difficulty with theories about practices. To take the second one first: the way in which our action has been muddled and self-defeating is usually not clear until we have made the transition and achieved greater insight. Now we are often (even always?) aware that our action is not fully coherent, but prior to the growth in insight we cannot pinpoint in what the incoherence consists. Perhaps it just comes from an inadequate application of theories we already master which bear on objects resistant to self-understanding. Perhaps, that is, there is no need for further insight into the very nature of our practices, their constitutive norms. Perhaps, indeed, we have already reached the limit of such insight. No-one expects that we will ever reach totally coherent, totally self-transparent action. Perhaps the existing muddle is part of the ineliminable remainder of confusion that attends human life at its best.

From the other side of the theoretical transition, it seems quite clear that there is a quite determinate muddle here which engenders an incoherence of a quite definite kind. But how does the theoretical pioneer convince the stay-at-home? Perhaps the greater clairvoyance of his practice is visible to outsiders, and that will convince them. But this is very often not so. Then the antagonists are left without a definite pin-pointable breakdown, that they can *both* recognize as such, around which the debate can turn.

We may grant that theoretical disputes in natural science can also run into this kind of problem. But they have other means available to them when it comes to continuing argument. In certain situations it can be in principle possible to identify a putative breakdown for both interlocutors; say, some experimental result which has turned out unexpectedly. Of course, the theorist who is incommoded by this can have recourse to all sorts of other hypotheses. This is a commonplace of Quinean wholism. No theory ever encounters a crucial experiment. But at least the debate can concentrate on clearly and inter-subjectively identifiable results, which as putative refutations serve as the focus of debate. In disputes between theories about practices, not even this focus is in common between the parties.

So in political theory, even when you are right, convincing your opponent can be very difficult, sometimes next to impossible. But how do you convince yourself? That is, how do you ensure against self-delusion, when you feel sure that your practice has become more clairvoyant?

In a way, this is an unfair, even an absurd question, as the sad history of modern epistemology has convinced us. There cannot be a criterion of non-self-delusion which would itself be proof against self-delusion. The ultra-sceptical question, how can you be sure? can be asked of any form of knowledge, even of the best established parts of natural science.

But there is nevertheless a point in raising this question here, because in the theories about practices, and particularly political theory, there are especially virulent forms of self-delusion which belong to the very type of thinking involved.

We can understand this better if we try to see why people adopt political theories. The stimulus to think theoretically about politics comes from the sense that our constitutive self-understanding is somehow inadequate. I mean by this the understanding implicit in our practices by virtue of their constitutive self-descriptions. But today just about everybody believes that to understand politics you have to have mastered the correct theory. Why is this?

In part, it is because of the rise and prominence of political economy. We are all convinced that there are mechanisms of social interaction which are not clear on the surface, regularities which have to be identified through study and research. Even people who are not at all uneasy about the implicit understanding of the society's institutions, and are not at all tempted to think that this understanding is somehow illusory, nevertheless accept that there is more to social interaction than can meet the eye. There are laws of soceity which have to be laid bare in a theory.

But people also turn to political theory because they feel the need to get clearer what society's practices involve. These practices seem problematic because they are already the locus of strife and trouble and uncertainty, and have been since their inception. I am thinking in particular of the central political practices of modern western democracies: elections, decisions by majority vote, adversary negotiations, the claiming and according of rights, and the like. These practices have grown in our civilization in a context of strife, replacing, sometimes violent, earlier practices which were incompatible with them. Moreover they are practices which by their nature leave scope for struggle between different conceptions, policies, ambitions. By their very nature, they are not practices which are likely to be the object of a tranquil implicit consensus, at least not for long.

VI

And so our society is a very theory-prone one. A great deal of our political life is related to theories. The political struggle is often seen as between rival theories, the programmes of governments are justified by theories, and so on. There never has been an age so theory-drenched as ours.

In this situation, while political agents may turn to theories as guides, or as rhetorical devices of struggle, many others turn to them in order to orient themselves. People reach for theories in order to make sense of a political universe which is full of conflict and rival interpretations, and which moreover everyone agrees is partly opaque. When in addition, people's purposes are frustrated in unexpected ways, for example, when they are beset with intractable stagflation, or anomic violence, or economic decline, the sense of bewilderment is all the greater; and the only cure for bewilderment seems to be correct theory.

In these circumstances, the scope for self-delusion is particularly wide. Whether people turn to theories as instruments of struggle, or as points of orientation in a confusing world, they are not likely to be easily capable of viewing them dispassionately. This will be especially evident when we reflect that people seek orientation in their political world not just to have a cognitively tidy universe, but much more because they need the political realm to be a locus of important significance. Either they want political structures to reflect their central values, or they require that political leaders be paradigms of these values, or they seek a form of political action which will be truly significant, or they require the political system to be the guardian of the right order of things; be it one way or another, they are reluctant to look on political structures simply as instruments which are without value in themselves — albeit an influential strand of modern political theory tends in just this direction.

But then the very satisfaction of becoming oriented, and feeling related to something significant, may give one a quite specious sense of having achieved more clairvoyant practice. This can generate very powerful mechanisms of self-delusion.

Everybody no doubt has their favourite examples. I will offer two that seem to me clear cases of self-feeding delusions. The first is the theory of certain radicals, including some Marxists, who make a critique of 'bourgeois' democratic institutions and look forward to their supersession by some truer democratic system of councils of the revolutionary proletariat, or some other such figment of red imagination. The delusion lies in thinking that one can overcome the problems of representation, indirect control, apathy, cross-purposes, threatening elitism, any more effectively through the imagined institutions than one can through the existing ones. But when people come to believe this kind of thing two consequences may ensue. First, the existing 'bourgeois' institutions may easily be made even less responsive to people's demands and aspirations, to the extent that they are disrupted by dissidence. And second, the revolutionary dissidents themselves may build a counter-community, in which there is a strong common purpose, and sense of oneness and comradeship; in short the very things that a self-governing community ideally has, and that the larger society lacks. The contrast between the faltering social institutions, and the successful practice of intra-party community, may provide all the 'proof' that seems to be needed; it may seem a paradigm case of gain in clairvoyant practice.

This, of course, it is not; because the theory is about the self-governing practice of whole societies, and the positive experience is in the life of a sect. But the imagination can leap this gap; it can picture the whole society with the intense unity of a sect. The theory can seem confirmed; it is entrenched by the very intense satisfactions of a community organized around a highly significant purpose.

Another delusion, symmetrically situated on the right, is a view of society made up of generally self-reliant individuals, whose common institutions have as function to maintain order and to offer support in those special cases where self-reliance is inadequate to the task. The view generally held by such people is that government these days has invaded our lives far too much, has taken on too great a responsibility, and has to be 'rolled back'. Public expenditure has to be drastically cut; public intervention ended in a host of areas, etc.

In their view, the good life is one in which the individual consumer has the maximum disposable income, and the greatest freedom to use it as he sees fit. But of course one of the chief causes of rising infra-structural costs in modern society is exactly this affluent individual consumer way of life. Affluence with a high degree of individual mobility means large conurbations combined with motor cars, a way of life which involves fantastic cost in highways, throughways, urban reconstruction, suburban sprawl, garbage disposal, use of energy, and so on. Or again, the highly mobile nuclear family cannot but require public authority to take over the burden of dealing with the old, the chronically sick, and so on.

The individual-consumer life-style runs up huge costs which can only be met by government, but at the same time it breeds the delusive consciousness that government is somehow taking away our money for purposes utterly foreign to this mode of life. This outlook dissociates the two sides of a coin; delusions do not get deeper than that.

And yet the delusion can be very stable, and for reasons rather parallel to the radical one. First, there is an element of the self-fulfilling prophecy. A public sector which is mainly perceived as a costly mistake will probably maximally approximate to a costly mistake, even though it remains essential, and is thus irremovable. But by remaining something of a dog's breakfast, it seems to justify all the strictures of the theory.

And second, there are tremendous satisfactions in this delusion. One reaffirms one's own sense of dignity as a self-reliant individual, capable of exercising responsibility for one's own fate, if only the all-invasive government, in collusion with the weak and irresponsible, did not get in the way. This identity of self-reliance is for a variety of reasons, which I have not time to enter into here, very important in modern western society. Hence atomist theory generates satisfactions which can be taken for clairvoyant practice, but which are really the height of self-defeating muddle.

But this delusion is powerful. It is now electing governments, in this country and latterly in America. What is worse, it seems to have infected the governments themselves, so that they build their policies around it. The punishment wrought by the real world will be truly terrible.

Here, as in the first case, the only remedy is to focus attention on the practices the theory is meant to account for. Atomism is a theory about the functioning of society. Confronted with this, it fares lamentably. Where it makes us feel good is in the private sphere. As is often the case, the recourse against delusion is to force our gaze outward. But who can ever be sure that he has achieved that? Small wonder it is so hard to achieve a rational consensus in political theory.

Hard, but not impossible. At least I hope that the above argument has indicated what scope there is for rational criticism.

But rational criticism requires that we keep in mind the particular nature of political theory. And this is what I have been trying to identify in this essay. What I have been arguing is that political theories generally differ from theories in natural science in that they have a quite different relation to practice. These theories are in general about practices, with the result that they often by their very nature work to undermine or transform the practices. In this sense, although there are exceptions in areas like economics where the standard model is approached, social theories do not bear upon an independent object. The objects they bear upon are not resistant to the alterations in self-understanding which these theories bring.

This means that their validation follows a rather different pattern. It is not a matter of confronting them with independent facts, or only in part. They also show their validity or invalidity in the way that they transform practice. What makes our practice more clairvoyant is *pro tanto* valid theory.

But the increased clairvoyance of practice may be hard to demonstrate in intersubjective space. And it is not hard to delude oneself about it. Delusive theories can have their own satisfactions. We have constantly to focus our attention on the practices the theory is allegedly about. Only increased clairvoyance of *these* validates. On this basis, reason can assemble considerations for an eventual arbitration. But it is a slow and arduous process.

Accounts, Actions and Values:
Objectivity of Social Science

Amartya Sen

In this paper I examine the old issue of the objectivity of the social sciences. The crucial distinction on which most of the analysis presented here depends is that between (1) a statement or an account, and (2) actions related to that account, e.g., the action of giving that account, or that of doing the research leading to that account. I argue that the evaluation of the latter must be value-based in a way that the evaluation of the former need not be. In the light of that distinction some of the controversial issues are examined and explored.

I. The Need for Social Science

Factual statements on social matters are typically controversial — some remarkably so. But they are also — frequently enough — terribly important. It may not be *easy* to know whether these statements are correct, but very often we would certainly *like* to know whether they are. 'Britain is facing the most serious industrial crisis in its history', says the Leader of the Opposition.[1] 'There is no alternative', says the Prime Minister.[2] The Official Ulster Unionist M.P. for Down, South, warns 'of future conflict in Britain resulting from the apprehension of the indigenous population about the growth of an immigrant population with dual loyalties and

* For helpful comments and criticisms, I am most grateful to Jonathan Glover. I have also benefited from suggestions of Dieter Helm, Christopher Lloyd, and Charles Taylor.   [1] Reported in *The Observer*, 22 February 1981.
     [2] This has been a recurrent theme, and has been much repeated by the government; cf. *The Sunday Times*, 1 March 1981 (Hugo Young's column).

background'.[3] How correct are these remarks of Michael Foot, Margaret Thatcher, and Enoch Powell? All this may be controversial stuff, but it would be absurd to deny the importance of knowing whether or not these statements are correct. Much hangs on them.

At a more general level, there are other factual aspects of social matters that are important to resolve. Professor Halsey points towards 'a hierarchy of cultural capital' in modern society — Britain, America, Russia — which 'permits considerable mobility for a minority upwards and downwards in the hierarchy, but its main feature remains the continuity of familial status between generations'.[4] Is this view of mobility correct? At a still more general level, take this remark of Karl Marx: 'The more a ruling class is able to assimilate the foremost minds of a ruled class, the more stable and dangerous becomes its rule.'[5] Is this thesis of conservative consequences of selective mobility correct? It would be sad indeed if it were the case that we could never know whether these statements are true or false, justified claim or incorrect assessment.

There is clearly a *need* for social science, even if the term 'social science' — as Dr. Brus said in his lecture in this series — is shunned in Oxford.[6] The need is clearly there, but of course the existence of a need does not imply that it can, in fact, be met. Perhaps social science must remain an *unfulfilled* need? I would like to argue against that position, and assert the possibility of social science. We must see, first, what it is that we must expect of science in general and of social science in particular.

There is one specific issue that seems to crop up again and again when the possibility of social science is discussed. That is the problem of social science being 'value-loaded' — suffering from an incurable infection of what Gunnar Myrdal calls

[3] Reported in *The Times*, 27 February 1981.

[4] A. H. Halsey, *Change in British Society*, Oxford University Press, Oxford, 1978, p. 110.

[5] Karl Marx, *Capital*, Vol. III (1894); Foreign Language Publishing House, Moscow, 1959, p. 587.

[6] 'Marxism and Communism', Wolfson College Lecture, 24 February 1981; See below, chapter 7, p. 179.

'the political element'.[7] That issue will certainly require early attention.

II. Accounts and Actions

I begin with distinguishing between two different — though related — features of scientific work, viz., *statements* or *accounts* on the one hand, and *actions*, on the other. Statements assert things. Leaving out interpretational ambiguities, statements can even be seen independently of the authors. Actions, on the other hand, are peculiarly things done by particular persons, *or* groups. Various actions may be linked up with a statement. The act of stating — or giving an account — is only one of them. Asking the questions leading to that account is another. Researching on the questions is still another. Deciding to report is a further action. Publicizing the report is one more. There can be little doubt that these actions connected with science can be normatively examined just as actions of other types can be. We would get nowhere at all unless we distinguished between the value-features of these actions and the value-features of the statements themselves. There is also a distinction between the accuracy of a statement and the appropriateness of the action of stating it (or of publicizing it, or of doing the research leading to it).

Am I labouring the obvious? Perhaps so. But not noting this elementary distinction clearly has much to do with some of the claims regarding the impossibility of social science. Let me illustrate with Gunnar Myrdal's analysis of the problem. Myrdal started out as a no-nonsense positivist, looking for hidden values in allegedly scientific works — searching through the writings of David Ricardo, Robert Malthus, John Stuart Mill, Karl Marx, to modern writers, with the enthusiasm of a Customs Officer examining a row of doubtful suitcases. On finding all authors guilty, Myrdal proceeded first to outline how to do economic analysis without smuggling in values. This was in 1929. Twenty-four years later Myrdal went on

---

[7] G. Myrdal, *The Political Element in the Development of Economic Theory*, 1929; English translation by P. Streeten, Routledge, London, 1953.

record rejecting all that, moving from the pole of relentless pursuit of value-freeness to the other pole asserting the impossibility of scientific economic theory. Myrdal's position is worth studying since the popular thesis of the impossibility of social science derives much from the argument championed by the later Myrdal. Myrdal put his transformation thus:

> But throughout the book [the 1929 version] there lurks the ideas that when all metaphysical elements are radically cut away, a healthy body of positive economic theory will remain, which is althogether independent of valuations. . . . This implicit belief in the existence of a body of scientific knowledge acquired independently of all valuations is, as I now see it, naive empiricism. . . . There is an inescapable *a priori* element in all scientific work. Questions must be asked before answers can be given. The questions are an expression of our interest in the world, they are at bottom valuations.[8]

There is no distinction here between the question as to whether any economic theory can *stand independently* of values and the question as to whether any economic theory can be *acquired independently* of values. Myrdal answers the latter question in the negative and treats this as having resolved the former question also. But why? Clearly, scientific work — like other actions — will be influenced by the values of the people involved, leading, among other things, to the selection of questions, as Myrdal rightly points out. But that does not imply that theories thus acquired cannot be rejected or accepted independently of values that led to them, nor that the theories could not be simply true or false.

All this does not, of course, give any reason for thinking the contrary, to wit, that economic or social theories *can*, in fact, stand independently of values. That question has simply not yet been addressed. What has been asserted here is that the presence of value elements in the choice of *actions* leading to a social or economic theory — or for that matter any other kind of theory — must be distinguished from the presence of value elements in the *statements* of the theory. If such statements cannot, in fact, stand independently of values, then the reason for this must lie elsewhere.

[8] Ibid., p. vii.

While I reject Myrdal's argument, I think he is certainly right to emphasize the importance of values associated with scientific activities, and I must return to this question later. The importance of value elements in the *actions* connected with science does not derive from their alleged role in introducing value elements into the *statements*, but from the importance that these *actions* themselves have. We judge scientific achievement not only by the truth and correctness of the statements actually made, but also by the choice of questions, the priorities of research, the selection of findings, and the publicity given to conclusions. Scientific *action* — like all other actions — is open to normative scrutiny.

III. Statements: Interpretation and Beyond

I turn now to statements or accounts. There can, of course, be serious problems of interpretation. Statements that look the same may say quite different things. The background of actions can sometimes help to clarify the context of a statement. This does not, however, affect the criticism of not distinguishing between the two. Further, actions themselves may need interpretation, and exploring the context of a statement is not the same as examining the related actions.

'There is no alternative,' says the Prime Minister. For whom? For the country? For any government in power? For the present government? For the current Prime Minister? We cannot begin to assess the statement until we know what is being asserted.

Understanding the context is a deep and complex problem, and I do not propose to offer any cunning technique for achieving this instantly. I am concerned with assessing a statement *after* it has been understood, taking note of the context.

Suppose we have crossed this hurdle, and see what the statement asserts. We can now go on to ask at least two questions about it: (1) Is the statement true? (2) Is it a good statement? The first question sounds 'scientific', but the sceptics doubt that the social sciences are in a position to answer such questions. The second question makes the

'true' scientist blush: 'how do you mean "good statement"?'
Values are seen gushing out of that question. I take up the
question of truth first and that of goodness later.

IV. Knowledge and Experiments

Why can't social sciences answer the question of truth? Here
we meet a variety of arguments. Consider *causal* hypotheses
first. It is pointed out that doing experiments is not easy in
the social sciences. It certainly isn't.

It is, of course, very tempting to think that the most
natural explanation of the policy followed by the British
government in the last few years is the testing of monetarism
for the benefit of a handful of economic theorists, plunging
millions into unemployment for this scientific experiment.
Never has so much been owed by so few to so many!

Naturally, I repudiate this exciting suggestion altogether.
However, the fact remains that at great cost something has
indeed been learned, and while this was not an experiment,
some questions of truth have, in fact, been settled, including
(1) the inaccuracy of some types of monetarist predictions
based on a traditional measure of money in the form of the
so-called M3, and (2) the inability of the chosen monetary
instruments to act rapidly on the magnitude of M3. It is not
the case that causal questions must remain unsettled in the
absence of experiments.[9]

There are, of course, many causal questions in economics
that are difficult to settle. In studying relationships between
some variables a great many others are taken to be constant,
and the precise role of these other variables may be difficult
to disentangle. This is one reason — among others — as to
why economic theory has to be, as Sir John Hicks puts it, 'in
historical time'[10], the juxtaposition of these other variables

---

[9] Further, as Assar Lindbeck argues, 'the "death risk" for erroneous theories'
has in recent years increased considerably because of 'recent advances in testing
with the help of non-experimental data' *The Political Economy of the New Left*,
2nd edition, Harper, New York, 1977, p. 27.

[10] J. R. Hicks, *Causality in Economics*, Basil Blackwell, Oxford, 1979.

at that time specifying the background conditions for the interrelations to be studied.

There are indeed many questions of a causal type that have been satisfactorily answered within the specified historical context. Many others have not been settled convincingly, or at all, and quite a few have sharply conflicting results. The fact that several of the causal relations most important for policy fall in the last category certainly does limit the *usefulness* of· economics for policy making. But that complaint — well justified — is quite different from asserting that causal questions in economics, and in the other social sciences, can never be settled.

Further, not all factual questions in the social sciences are concerned with cause–effect relations. Some of them demand description of what is actually happening. I have tried to argue elsewhere that the usefulness of descriptive economics in this sense has tended to be systematically underestimated.[11]

The absence of experimental opportunity is less of a restriction here. Cosmology too makes do with what it can find, and cosmologists are reconciled to the life of the gatherer rather than of the hunter. There are observational experiments, of course, but exact parallels of that exist in the social sciences too.

V. Self-understanding and the Social Sciences

The real problem comes from elsewhere. In his contribution to this book,[12] Charles Taylor identified the basic questions of political theory as: 'What is going on? What is really happening in society?' He pointed out that 'a society is among other things a set of institutions and practices, and these cannot exist and be carried out without certain self-understandings'. And political theories 'transform this self-understanding, they undercut, bolster, or transform the constitutive features of practices'. 'In politics,' as Taylor puts it, 'accepting a theory can itself transform what that theory

---

[11] 'Description as Choice', *Oxford Economic Papers*, vol. 32, November 1980.
[12] Charles Taylor, 'Political Theory and Practice', above, Chapter 3.

bears upon'. The cosmologist does not have to face *that* problem.

What about the economist? It has to be mentioned straight-away that self-understanding is not one of the more popular themes in economics. The modern economist tends to be altogether too chary of discussing questions of such moment. But given the nature of the subject matter it is not easy to escape that issue, and Taylor-type questions do belong to economics.

Indeed, the issue has been faced in the past by several of the more visionary writers in economics. This certainly includes Marx, but perhaps more remarkably, the list includes that finicky old theorist Knut Wicksell, the great Swedish economist who is perhaps best remembered for his pioneering work on pure capital theory. Knut Wicksell saw political economy as having a radically transforming role on the subject matter of its study:

Among the eighteenth century Swedish writers on economics, . . . we repeatedly find remarks which show that the conception, so repellent to our minds, of a workman as a mere beast of burden was, as recently as two centuries ago, still general and deep-rooted. Indeed, it must be regarded in some degree as one of the merits of economic science that in this respect it has produced a revolution in public opinion. As soon as we begin seriously to regard economic phenomena *as a whole* and to seek for the conditions of the welfare of the whole, consideration for the interests of the proletariat must emerge; and from thence to the proclamation of *equal* rights is only a short step.
The very concept of political economy, therefore, or the existence of a science with such a name, implies, strictly speaking a thoroughly revolutionary programme.[13]

Gunnar Myrdal, who commented on this aspect of Wicksell's work, confessed to some bafflement, but offered a simple account of what Wicksell was up to.

. . . Wicksell's own conclusion is an obscurely worded plea for a *particular* political valuation; for his own view of what the goal should be for

---

[13] K. Wicksell, *Lectures on Political Economy*, Volume I ('General Theory'), translated by E. Classen, edited by L. Robbins, Routledge, London, 1934, p. 4.

social evolution. Wicksell's thought, elsewhere so lucid, becomes here extremely difficult to follow.[14]

But Wicksell's claim — right or wrong — was surely much more ambitious than the one identified by Myrdal — certainly not a 'plea' that others should accept Wicksell's 'own view of what the goal should be'. His claim surely was that the self-understanding that political economy generates through its treatment of economic phenomena as a whole and of social welfare makes it impossible to sustain the older view of worker and his place. And this shift in the conception of human beings must lead to revolutionary proclamations and programmes. Wicksell was presenting a thesis of interrelations — between cognition and valuation, between economics and politics, between the studier and the studied. This can hardly be seen as grafting his own political values on economic analysis that is otherwise taken to be neutral in its impact.

Was Wicksell right? It is a difficult question to answer. Certainly he over-simplified the picture and made very strong claims for political economy. Political and social changes taking place *pari passu* with developments in political economy must have affected the way political economy is interpreted. There is too little feedback in Wicksell's characterization and the influence is seen to be peculiarly uni-directional. Nevertheless, there is something of real substance and importance in Wicksell's analysis of this linkage. The conceptualizations of economic interdependences and of social welfare as a whole have, in fact, tended to push the broad currents of economic thinking increasingly in the direction of intervention and equality, even though 'revolutionary' is too flattering a description for this type of programme. The conceptions of priorities and rights have certainly shifted a great deal not only from the eighteenth-century characterizations, which Wicksell quoted, but also from the prevailing

---

[14] Myrdal, op. cit., p. xv. Myrdal is actually referring to the argument as presented by Wicksell in *Ekonomisk Tidksrift* (1904), but Myrdal points out that Wicksell 'never gave up this view' by quoting a part of the passage quoted here from *Lectures on Political Economy*.

views a hundred years ago. A comparison of the major themes in today's textbooks with those in textbooks around the turn of the century (e.g., in Marshall's *Principles*) shows the extent of the change of focus. Even the current challenges of President Reagan and Prime Minister Thatcher to 'roll back' government intervention, and the bitter complaints of economists like Friedman and Hayek, can be understood only in terms of this vastly changed background. The reorientation of political economy, which Wicksell identifies, certainly fits somewhere in this large story.

How does all this affect the status of political economy as social science? The answer depends upon what we take social science to be. If by social science we mean something very like natural science with the same techniques applied to social matters, then clearly all this about self-understanding and related matters as a part of social science is more than a bit dodgy. But that would be a peculiar way of characterizing social science, even though it is implicit in many writings. It is rather like defining aerial warfare as techniques of naval warfare applied in the air!

Social science is more sensibly seen as being concerned with advancing knowledge on social matters, taking into account the general *and* the peculiar features of social matters — self-understanding is one of the peculiarly social characteristics — and admitting whatever techniques would serve the goal of advancing knowledge. These techniques will often be quite different from those used in the natural sciences, which themselves admit considerable variety. There will, of course, be points of similarity also between natural and social sciences, and these points are of interest, but *not* for the alleged status and legitimacy that they bestow on the social sciences.

VI. Insufficiency of Truth as a Criterion

I had postponed discussing the question of the *goodness* of a scientific statment as opposed to its *truth*. I take up that issue now. First a preliminary query. Why do we need to make these goodness judgements at all? Why not just settle for the truth, as some purist scientists boast that they do?

This can, however, scarcely be adequate. What Isaac Newton called 'the great ocean of truth' contains all kinds of things — from the indispensable to the useless. We cannot, thus, escape judging. A description can be true but quite uninteresting. So can a causal theory. As an example, let me offer now a brand new piece of economic theory, which may be called the 'Iron Law of Doles'. It says that in a country with unemployment benefits, the size of dole payments increases monotonically as the volume of actual unemployment increases. I think this iron law will pass the test of truth. But it is most unlikely to be regarded as a memorable achievement.

One of the recurrent themes underlying the logic of current government policy has been: 'You cannot spend more than you earn'. In a literal sense this is, of course, false, but there is a sense — involving long-run balancing — in which this is a true statement. Is it a good statement? As an explanation of government strategy, I would say not at all. Critics of government policy — including Keynesians but others too — believe that a different policy would lead to more output and employment and, thus, more real earning. Any argument that specifies what you can spend *given* the earning is, therefore, no response to the criticism at all. Worse, in pushing the enquiry away from the important issue of variability of real earning, it hinders rather than helps the cause of understanding what really constrains the British economy. The statement remains true, but scarcely a good statement for enlightening or communicating.

VII. A Statement as an Account of Something

I turn now to the more difficult problem of the *nature* of judging the goodness of a scientific statement. What kind of a quality is goodness in this context?

The truth of a statement is, in an obvious sense, a characteristic of that statement itself, but the goodness of a statement — on any sensible interpretation — must depend also upon the purpose that the statement is supposed to serve. The same statement, without any ambiguity of interpretation, can be judged as an account of one phenomenon, or alternatively of another. The issue is not really to judge whether the statement is good *simpliciter*, but whether it is good *as an account* of something or other. Isaac Newton's statement, which I referred to earlier, 'I seem to have been only like a boy playing on the sea-shore, . . . whilst the great ocean of truth lay all undiscovered before me,' is clearly a better description of truth than of Newton, although Newton is ostensibly the subject of the statement.

In judging a statement in the social sciences, it can be terribly important to ask: what is it an account of? Take a statement specifying the mean consumption basket of the British consumer, stating the amounts of each type of good consumed per head. As a description of the variety of goods that the British public as a whole does buy, this might be quite a good statement, but — at the same time — as a statement of the standard of living in Britain this is clearly quite a bad statement, if only because it is blind to distribution.[15] Goodness of the same statement can vary depending upon what we take it to be an account of.

VIII. Good Account to Give

This interpretation of the goodness of a *statement* has to be distinguished from the merits of the action of *making* the statement, or the action of *using* it as an assumption. A statement that is not a good account of something might nevertheless be a good statement to make — even as an account of that thing — on some other grounds. Your neighbour has gone off on a holiday, but when a suspicious-looking character asks you about this, you put on your burglary-prevention hat and say that he hasn't. I can never judge these things, but let me accept that this was a good account *to give* as to where your neighbour was. But it certainly was not a good *account of* where your neighbour was.

This type of distinction is simple enough, but overlooking

---

[15] The links between distributional judgements and national income evaluation are explored in my paper, 'Real National Income', *Review of Economic Studies,* vol. 43, 1976.

it has caused some confusion in methodological debates in economics. In the famous battle between Paul Samuelson and Milton Friedman on the case for using unrealistic assumptions in predictive economic models,[16] Friedman argued powerfully and convincingly for the use of false assumptions if they lead to good predictions. False assumptions might provide simpler and more usable models, and if they predict well, the falsity of the assumptions need not be held against the model. So far so good. But then Friedman proceeded to argue that the person seeking realism must also take such untrue statements as *more realistic*, using 'the test by prediction' to 'classify alternative assumptions as more or less realistic'.[17] This seems to confound two quite distinct issues. While a false statement about something might well be a good description *to give* on some instrumental grounds — in this case for the particular task of better predictions — it could scarcely be taken to be a 'realistic' description *for that reason.*[18] A bad account of something does not become a good account of it just because it has been found to be a good account *to give* on instrumental grounds. You might have been right to say that your neighbour *wasn't* on holiday — a good description to give — but that did not make it a *realistic* description of where your neighbour really was, or a good account of it.

IX. Use-interests, Conditionality, and Truth

In choosing rather simple examples I may have been over-simplifying the nature of the judgement of a statement as a good account of something. In fact, in the particular example

[16] M. Friedman, *Essays in Positive Economics*, Chicago University Press, Chicago, 1953; P. A. Samuelson, 'Problems of Methodology: Discussion', *American Economic Review*, vol. 53, 1963, reprinted in J. Stiglitz (ed.), *The Collected Scientific Papers of Paul A. Samuelson*, M.I.T. Press, Cambridge, Mass., 1966. See also S. Wong, 'The "F-twist" and the Methodology of Paul Samuelson', *American Economic Review*, vol. 63, 1973; L. A. Boland, 'A Critique of Friedman's Critics', *Journal of Economic Literature*, vol. 17, 1979.

[17] Friedman, op. cti., p. 33.

[18] This issue and the Friedman–Samuelson controversy in general are examined in greater detail in my 'Description as Choice', *Oxford Econimic Papers*, vol. 32, November 1980; reprinted in my *Choice, Welfare and Measurement*, Blackwell, Oxford, 1982.

concerning your neighbour, truth alone turned out to be sufficient to determine what could or could not be taken to be a good statement of where your neighbour was. In a complex case, truth alone, or realism more broadly defined, may fail to be so decisive. Let me enrich the example. Your neighbour's daughter has developed symptoms of flu at school, and the headmistress asks you on the phone whether you can tell her where your neighbour is. You are a practised hand by now at answering that question, and you respond instantly: 'He isn't on holiday.' Not very useful, the headmistress thinks, and asks for more detail. Is he in the factory where he works? Is he out shopping, and if so where? Clearly, the goodness of the description of where your neighbour is has to be judged taking into account the purpose for which this information is being sought. At the risk of over-simplification, call this the 'use-interest' of the account.

On this approach, the goodness of a factual statement is judged by specifying both the thing of which this is taken as an account and the use-interest of the information in terms of which this evaluation is to be made. We are really asking, then: *Is this a good statement judged as an account of x from the point of view of use-interest* y? Consider the variety of use-interests that might be involved in seeking information on where your neighbour is. To give just a few examples: (1) for tracing him to tell him about his daughter; (2) for planning a neat little burglary in his house; (3) for deciding whether milk bottles should be left for him outside his door; (4) for planning a riotously noisy party that night; (5) for a spot of gossip as to whether there is any connection with the woman two doors away also going off on a holiday.

These are, of course, trivial examples, but exactly similar pluralities of use-interests can exist about factual statements on graver social matters. Whether we are examining national income, inequality, poverty, unemployment, or cost of living, *or* forecasting future economic peformance, *or* examining causal links between money supply and unemployment, between closed shops and wage bargains, the question of use-interest introduces a strong element of conditionaltiy in judging these statements.

A suspicion might arise here. I have been treating use-interest as a *supplement* to truth as a criterion of judgement. But can't use-interest *supplant* truth itself? What if the user is not interested in the truth, but just the opposite? Isn't supplementation by use-interest the thin end of the wedge, leading ultimately to banishment of truth itself, and along with that, all links with science altogether? There is, I think, need for clarity here.

Any use-interest in the information provided by an account must obviously be concerned with information as such. A person might well not want that information and may want instead the speech-act of consolation, say, from his doctor. The interests of a person *as a person* — taking everything into account — have to be distinguished from the use-interest of an information, irrespective of whether or not that use-interest relates to the person. If a use-interest of this type does not apply to the person in question, then that's fine, he is not involved in this exercise. The fact that he also wants consolation is a separate story, like wanting love, or cuddling, or excitement. That story does affect, of course, the doctor's choice of *actions*, including choice of statements that he should make, but all this is on a different plane altogether from judging the information given in a statement from the point of view of use-interest of that information. Use-interests are essentially impersonal, even if they are articulated in a personal form. I shall have to return to this issue again while examining the question of objectivity, which I shall take up presently.

X. A Good Account vis-à-vis a Value-Loaded Account

Let me begin, once again, with a very simple question. If accounts in the social sciences can be good or bad, does this not compromise the objectivity of those allegedly scientific statements? Does it not make the accounts value-loaded, just as Gunnar Myrdal thought?

But why? We must distinguish between a statement or an account being value-loaded and a judgement *about* that statement or account being value-loaded. If I were to look around and say what a good building this one was, you might think that my judgement was value-loaded, but you are unlikely to think that the building *itself* was value-loaded. The distinction here is transparent enough, since value-loading is not the kind of thing that happens to buildings — at least none is known to have collapsed because of excessive loading of value. A statement or an account on social matters — unlike a building — *can* conceivably be value-loaded, but comparably with the case of judging the building, the statement or account cannot be shown to be value-loaded by the device of pointing out that a judgement *about it* (viz., that it is a good statement or account) is value-loaded.

That was a small hurdle to cross. And it does, of course, leave open the possibility that judgements of scientific achievements and thus scientific *standards* must be value-loaded, since they are based on judgements of goodness of accounts. In one sense, this seems inescapable: how can a judgement of goodness (in this case, of accounts) not be value-based? Does this compromise the objectivity of social science by making its standards value-based? I shall argue that it does not.

XI. Objectivity and Non-uniform Evaluation

First, a preliminary remark on this challenge to the objectivity of social science. This problem — in this general form — cannot be a special problem of social science only. Presumably judgements about goodness of accounts and theories in natural sciences must also share the same predicament, whatever it is. What makes an account in the natural sciences good is also, in the same sense, a matter of values. There are, of course, many special problems for the social sciences without parallel in the natural sciences, but here we are not looking at one.

More importantly, insofar as statements or accounts are judged to be good from a specified point of view, viz., that of a particular use-interest, the judgement is rather like describing a knife to be a good knife to cut bread. The value element is not fundamental here, since the goodness is seen as an instrumental merit.

This conditionality vis-à-vis use interest may raise other worries. Isn't there a danger of 'relativism' here? If different people can adopt different points of view, isn't the whole thing fundamentally subjective? Isn't this the end of the objectivity of scientific assessment?

I believe that this fear is unwarranted; the question of objectivity does not turn only on the existence of relativity as such but also on what it is relative to. If one were to say, 'I feel this is a better account and that's all there is to it', then the nature of the judgements must be — in some obvious sense — subjective. But if the differences between different people's evaluations are due to contrasts of the use-interests in terms of which the accounts are judged, then what we have is relativity vis-à-vis different objective phenomena. Lack of objectivity might well lead to relativity, but relativity does not imply lack of objectivity.

One difference between natural and social sciences might lie in the greater homogeneity of use-interests in the natural sciences compared with social. This makes standards and judgements relatively more uniform in the natural sciences than in the social sciences. But this problem of uniformity should not be confused with that of objectivity, since objectivity is consistent with non-uniform evaluation when the variations are related to objective differences of use-interests.

An analogy might help. *Seas* are compared in size almost always in terms of surface area. Different *countries* are sometimes compared in terms of surface area (the Soviet Union is the largest), and sometimes in terms of population numbers (China is the largest). There is no difference in the *objectivity* of comparing sea sizes vis-à-vis comparing country sizes, despite the plurality of standards in the latter. What is needed is clarity about the nature of the exercise that is being done.

XII. Actions, Science, and Values

I move now from statements and accounts to actions. If the role of values is typically overestimated in the case of scientific accounts, they are frequently underestimated when it comes to actions. The problem here isn't fearing that scientific action *might be* value-loaded, but fearing that it *might not.* Value-loading here is not so much a right as a duty. An action by a person that is contrary to his or her values — taking everything into account — remains pernicious in terms of his or her own values, even if it happens to be related to science.

Of course, an expert might justifiably give advice — say to a government — on the means of pursuit of goals that are not his own, if, on balance, all things considered, he does not find this course to be morally wrong. But the moral question has to be posed and the answer taken seriously, like in any other conscious choice, and the fact that one is engaged in science or pursuing the truth is not an argument for exempting one-self from this discipline. My point is that actions related to science are like all other actions, calling for evaluation, assessment, and scrutiny.

And this applies to *every* aspect of scientific activity: selecting fields and questions, doing research, gathering results, reporting on them, publicizing findings, and everything else. Here again, there is no qualitative difference between natural and social sciences.

XIII. An Illustration

The different natures of judgements of statements and actions can be illustrated with a concrete case of scientific activity. Take the controversial Panorama programme of the BBC on brain death and kidney donations shown in October 1980. It may be recalled that this programme cast doubt on the certainty of death of allegedly dead patients when their kidneys were removed, and this show led to a tremendous decline in kidney offers. As a Consultant in the Renal Unit of Manchester Royal Infirmary put it: 'In terms of kidney transplantation, this is like the Wall Street crash!'[19]

There are two distinct evaluative questions about this case of scientific reporting:

(1) Was it a good *account* of brain death and kidney removal?
(2) Was the *action* of broadcasting the report right?

I am not really concerned here with the substantive problem of answering questions (1) and (2), but with using them as pegs on which to hang some methodological points. The report has been described by many medical practitioners as inaccurate, and certainly it misrepresented British practice in its summary assessment. On the other hand, it did draw attention to a potentially important problem that has arisen elsewhere and could arise in Britain unless the medical practices are more closely scrutinized.

As argued earlier, the first question can be answered only in terms of specific use-interests. It is, of course, quite possible that some use-interests were badly served, e.g., those of potential kidney-donors who were mislead about British practice, made unnecessarily scared, or induced to take decisions on unrepresentative information. Others, however, were probably well served, e.g., the use-interests related to medical safe-guarding, if the concerned doctors were helped by the information — unrepresentative though it was — to reconsider the existing safeguards against the kind of problem described in the programme. Two points now.

First, it is tempting to ask: Was the report a good account *everything considered*? But what is the point of asking that question? If the intention is to decide whether to *give* that account, then what is in the dock is the *action* of giving that account, and not the *account* itself. On the other hand, if the intention is to check whether the account is 'useful', then the answer depends upon what particular use-interests are taken as the basis of assessment. It was useful from one point of view, but not another. The 'everything considered' judgement applied to the account itself is otiose.

[19] Reported in *The Times*, 5 February 1981.

Second, not every group of people affected by the broad-casting of the programme may have a use-interest in the report. Consider the renal patients. Clearly, their well-being *was* affected by the report. But that is not the same thing as saying that their *use-interest in the information* was badly served. Insofar as they were curious about medical practice and were misled by the programme, their use-interests would have indeed been badly served. But this has nothing to do with their well-being — and in a different sense their 'interest' — being ill-served by the fall in kidney donations as a result of the programme. As renal patients they had no use-interest in the *information*. They were interested in getting kidneys, and that practical interest was ill-served, but no use-interest in the information is involved in this relation. The kidney patients are simply not involved in the judgement of the scientific account (except as members of the public concerned about medical practice).

When, however, we move from question (1) to (2), the kidney patients must loom large in judging the rightness of the BBC's action in broadcasting the programme. The fact that these patients were not interested in the information as such does not, in any way, rule them out from this judgement of *action*, as opposed to that of the goodness of the *account*. All the people affected by the decision have to be considered in this action judgement.[20]

Furthermore, the action judgement calls for an answer as to whether it was right for the BBC to broadcast that programme everything considered. Should it have done it? There is, however, no corresponding question in judging the programme as an account. 'Should the BBC have given such an account?' is a question about action judgement, not about account judgement.

Account judgements and action judgements are, thus, fundamentally different. The former will be use-interest relative, but the latter can take on the 'everything considered'

---

[20] For an illuminating analysis of problems arising from such conflicts of interests, see G. Calabresi and P. Bobbitt, *Tragic Choices*, Norton, New York, 1978.

form. The former would rule out interested parties whose interest lies in matters other than the information presented in the account. Things have to be considered in the action judgement that have no place at all in the judgement of the report as an account.

XIV. A Concluding Remark

Moral judgements in the fields of science — social *and* natural — directly concern actions, not accounts. This is because action judgements of *giving* an account have to face conflicts, take sides, or arbitrate. The same applies to *choosing* fields of research, *selecting* questions for investigation, and *picking* the ways of presentation. However, in judging a scientific *account* itself, there is no requirement to arrive at a 'comprehensive' judgement taking note of the different use-interests and their possibly conflicting demands. The judgements can be relative to use-interests, without having to judge the use-interests themselves.

The need for a combined judgement is a peculiar compulsion of choice of *action*. We have to do one thing or another; or nothing; but we have to choose. This applies just as much to actions specifically related to science. For *accounts*, on the other hand, the 'comprehensive' judgement is quite gratuitous. There is no necessity there, as there *is* in the case of actions.

The objectivity of science is not contradicted by the value basis of the scientist's actions. At the same time, the fundamental need for values in the choice of every *action* of the scientist — without exception — is not avoided by the objectivity of science.

Social Theory, Social Understanding, and Political Action

John Dunn

I

Every human being who is not in some way fundamentally cognitively damaged is at least an amateur social theorist. To acquire a human culture at all and to perform human actions both require social understanding. Being human is not something which we simply *are* but rather something which we learn; and much of learning it is learning how to understand the social relations in which we find ourselves. Because this is such a central truth for all human beings, it is not merely the case for each of us at an individual level that we are perforce amateur social theorists to a person, but also true at a social level that every human society embodies a more or less coherent 'official' social theory or set of social theories. But the term 'social theory' in modern speech, of course, does not in general refer to these relatively universal commonplaces of official doctrine or of the grammar of social membership. Instead it points, a trifle waveringly to be sure, to something more distinctive about the social relations of today, in Britain or in Zaire or in Chile or in North Korea. Social theory is what we have with which to close the gap between our individual social understandings and our experience of modern history, to reconcile, perhaps, the real with the rational. For us it is the intelligibility of what we know to be a largely unintelligible human world, the acceptability of

*I am very grateful to Susan James for her thoughtful criticisms of the initial text of this lecture, particularly since lack of space and incapacity have prevented me from meeting so many of them.

an always potentially intolerable condition, a true opium of the intellectuals. Like opium, however, it is not merely addictive — in that we have come to rely more and more upon it, to find it less and less dispensable — but also diminishingly effective, requiring ever larger doses to sustain its desired impact, and increasingly disastrous in its side-effects.

Expressed in these terms, this may appear a somewhat hysterical judgement. But it is not a judgement which is at all difficult to defend. What I attempt here, in crude outline, in order to make the judgement seem a trifle more real is, firstly, to explain how this condition of ideological intoxication has arisen; secondly, to suggest how we could realistically seek at least to initiate the slow process of detoxification; and thirdly, to indicate how directly and how painfully these issues obtrude in the present condition of the politics of Britain.

II

To take the explanation first: it will be apparent at once that this is a complicated matter. At one level, in order to explain at all adequately the place occupied by social theory in modern thought and modern experience, it would be necessary to tell, more or less concisely, most of the history of the human race. But being less ambitious, and simplifying wildly for the sake of intelligibility, we can see it in the first instance simply as a direct product of the secularization of modern culture. Within an explicitly secular culture the actual social understandings of the individual members of a society can in principle be joined onto the official social theory or doctrine of the society to which they belong only by an elaborate normative theory and an apparatus of causal beliefs. The history of modern social thought has seen extensive squabbling about the theoretical terms of trade in this conjunction. Methodological individualists and exponents of some varieties of utilitarianism, for example, have seen the sensory states and the actual social understandings of individual human beings at a particular time as the sole locus of human reality at that time and treated the social relations in which individual human beings find themselves simply as inventories of resources to be exploited by each of us, on the basis of optimal causal understanding — as effectively as possible — for our personal gratification. This is not a distinctively modern view. Indeed, eloquent statements of it can be found among the thinkers of ancient Greece. But there is an extremely strong historical relation between it and the self-understanding of capitalist economies; and in historical perspective, the vulgarization and dispersion of this view throughout the population of modern capitalist societies is by now as decisively a characteristic of these societies as the levels of material consumption which have been attained within them. A culture, in this view, is simply something which individual human beings happen to acquire — and to do so more or less thoroughly, depending upon the contingencies of reinforcement. Methodological wholists, by contrast, firmly reverse the perspective, seeing culture (within a variety of causal frames) as constitutive of human reality (as the site of being human at all) and seeing individual human beings accordingly as instances which cultures happen more or less thoroughly to acquire.

The fact of socialization, of course, and its real, if limited, efficacy in the case of most human beings, is not something which any serious social theorist could have sound reason to ignore. But it is not a fact which bears its own meaning on its face. It is not in any sense self-interpreting. To appear susceptible to socialization, to seem its willing victim, is on the whole a prudent investment for anyone. But to be too susceptible to it in actuality might well prove an existential disaster. Empathy may be a good servant to the ego. But it may also be a harsh master. A subtle attention to the expectations of others and an insight into their feelings may, in the cool and self-controlled, be a considerable aid to getting one's own way. But a degree of moral self-repression and an incapacity to avoid feeling for the feelings of others can be a nasty obstruction to realizing one's own desires. How does it really make sense for human beings to live?

In an explicitly religious culture this is a question which

could at least anticipate an answer drawn from outside the space of human social existence altogether. In a secular culture, however, if it is indeed a question which can expect any answer at all, that answer must certainly be drawn from inside the space of human social existence. For most of us today, the space of human social relations is all the evaluative space there is. The answer might not, of course, draw its central theoretical term directly from *social* experience as such. Utilitarianism for example, at least in some versions, puts its trust as directly as it can in materially given sensory states. But even the most crudely hedonic versions of utilitarianism sanction the promotion of the highest aggregate of preferred sensory states, however distributed, which they deem to be socially possible. Social possibility is a real causal constraint on the sensory states in which human beings can in practice find themselves. This is not the context in which to assess utilitarianism as a theory of the human good. But even if utilitarianism is theoretically determinate in principle (which I doubt) and even if it can be related convincingly to the theory of individual practical reason, it is hard to see how it can be *read* clearly at all as a *social* theory, a theory of the relationship between individual existence and social reality. Within the constraints of a particular culture, no doubt, it may serve well enough as an idiom of loosely prudential advice. Consequences certainly do matter and it is as well to watch out for them as best one can. But as a theory of what sort of culture human beings have good reason to seek to fashion for themselves and their descendants, or as a theory of what sort of human beings it is desirable in principle for cultures to contrive to form, it is extremely hard to believe that utilitarianism in fact offers any clear direction at all apart from a certain mild disparagement of what David Hume called the 'monkish virtues'.[1]

[1] David Hume, *An Enquiry concerning the Principles of Morals*, IX Pt 1, 219 (*Enquiries*, ed. L. A. Selby-Bigge, Clarendon Press, Oxford, 2nd edn. 1902, p. 270). 'Celibacy, fasting, penance, mortification, self-denial, humility, silence, solitude, and the whole train of monkish virtues; for what reason are they everywhere rejected by men of sense, but because they serve to no manner of purpose; neither

But if the answer which utilitarianism offers to the question of how it really makes sense for human beings to live is a trifle feeble, because of its virtual silence on the nature of social relations, at least utilitarianism does possess an answer of sorts to the question. It is far less clear that the same holds true for its most important modern competitor, Marxism. As a conception of how we have good reason to live, utilitarianism is theoretically sincere but humanly unconvincing. (No doubt each of us enjoys the pleasures which please us; and we certainly prefer our own preferences. But are either of these really very deeply directive thoughts?) Marxism, by contrast, at least within capitalist societies, is often humanly rather compelling on the matter of which side it is more creditable to be on in a particular brawl (or at the very least which side it is better not to be on). But on the issue of how human beings do have good reason to live, Marxism is both deeply incoherent in its fundamental tastes and profoundly evasive in its tactical pronouncements. ('Be good enough to leave that question till later. No decent person would ask it now' — and so on. And later, when it *is* too late, 'Well, that's all just blood under the bridge now, isn't it?') Marxism may be cogent enough at times on the moral and practical implications of the immediate social context of action. But as a comprehensive conception of how human beings do have good reason to live their lives, it is theoretically disingenous to a truly staggering degree. Now it is possible (though not in my view correct) to dispute the claim that Marx himself ever conceived his own views as a comprehensive conception of how human beings have good reason to live; and it is relatively simple to acknowledge that it would have been intellectually more prudent in any case for him to have repudiated any such ambition. But what cannot plausibly be denied is that

advance a man's fortune in the world, nor render him a more valuable member of society; neither qualify him for the entertainment of company, nor increase his power of self-enjoyment? We observe, on the contrary, that they cross all these desirable ends; stupify the understanding and harden the heart, obscure the fancy and sour the temper. . . . A gloomy, hare-brained enthusiast, after his death, may have a place in the calendar; but will scarcely ever be admitted, when alive, into intimacy and society, except by those who are as delirious and dismal as himself.'

the claims to political authority of Marxism as a theory of practice are deeply involved with this essentially spurious semblance of comprehensiveness in scope, and that the full menace of this intellectual hubris can be legitimately read in the deformations which it has undergone in its modern historical career as a doctrine of state. At the very least the terms in which Marx lived his own intellectual life gave, from the point of view of his subsequent auditors, altogether too many hostages to fortune.

It is, of course, a familiar enough theme in conservative polemic to insist that Marxism has sacrificed its promise as an analytic approach to the understanding of society and history to a superstitious and morally illicit claim to political authority and political power. But, although there is a good deal of truth in this conservative judgement, it is not my wish here simply to reiterate it. Indeed, I should like instead to suggest that precisely such an elision between analytic understanding and political pretension is a standing peril for all modern social theories. It is the single key characteristic of any serious modern social theory as such that it must in this manner seek to close the gap between our individual social understandings and the official doctrines of particular societies, affirming, negating, qualifying, or simply interpreting the latter. To close this gap would be to stretch across a space which there is no guarantee in principle that human imagination and understanding can in fact bridge. Insofar as the gap is in fact closed by veridical understanding in a particular society at a particular time, this analytic insight is in no way guaranteed to retain such a status in any other setting and at any other time. In a sense, most of modern social theory, where it does not simply consist of the more or less dreary reiteration of devout tautologies, is predominantly a forlorn hoarding of scraps of past understanding, reminiscences of fugitive episodes or passages of comprehension, recollected sometimes in tranquillity but often in rising anxiety: in Thomas Hobbes's phrase, 'decaying sense'[2]. To be sceptical

---

[2] Thomas Hobbes, *Leviathan,* ed. Michael Oakeshott, Basil Blackwell, Oxford,

about the truth status of modern social theories is not an exacting intellectual project. Considered as technical devices for understanding the reality of the societies in which we live and the extraordinarily complicated causal ecology within which these societies subsist, the full resources of modern social theories (despite in some cases their formidable intellectual intricacy) are no more than a pitiful array of intellectual knick-knacks.

But, if to feel authentic scorn for the achievements of modern social theory is a pretty effortless feat, a mood of more or less complacent personal contempt is hardly today a satisfactory surrogate for a theory of society. The place of social theory in the material reality of modern societies is not simply a random mishap. If there is one thing which is as certain about our lives today as the intellectual impossibility of taking modern social theory wholly seriously, it is the absurdity of hoping simply to dispense with it.

The ideological synthesis of political authority or of the subjective plausibility of collective political projects in modern societies (Great Britain, Poland, Uganda) is certainly both intellectually and morally a perilous venture. (When was it ever not?) But if such synthesis is not even attempted in intellectual and moral good faith, it will simply be substituted for by even larger increments of pure violence than those to which we have already become accustomed. If a measure of social integration and compliance is not secured in societies today through opinion, it will simply be imposed by force, since the conditions in which we live today are practically unacceptable to almost all of us without such a measure of social integration and since, where this minimum is indeed under active threat from inside a society, anyone who can

1946, Pt I, Ch. II, pp. 9, 10. 'IMAGINATION therefore is nothing but *decaying sense*; and is found in men, and many other living creatures, as well sleeping, as waking . . . This *decaying sense*, when we would express the thing itself, I mean *fancy* itself, we call *imagination*, as I said before: but when we would express the decay, and signify that the sense is fading, old, and past, it is called *memory*. So that imagination and memory are but one thing, which for divers considerations hath divers names.'

muster sufficient force to guarantee it will also muster a minimum of political consent. The reasons why the Argentine is still governable today and why the Soviet Union is still governed by the Communist Party have more in common than would be likely to appeal to the political leaderships of either country.

The choice which is open to modern societies — and, more importantly, the choice which is widely *known* to be open to modern societies — is a choice between the extent to which their social integration rests simply on opinion and the extent to which it rests directly on force. There are no explicit social theories of any importance extant today which happily endorse the resting of social integration solely on force (and perhaps, under reasonably close inspection, there have not been many such theories in the past either).[3] But if social integration today is widely recognized as materially necessary and if, at least in secular societies, it is also expected ideally to rest its weight on secular opinion, one key question in modern social theory is how far such opinion can be sanely hoped to consist of true beliefs and how far it must be resignedly expected to consist of beliefs which are in fact false. Now this, of course, is a very old question in political theory and one which has on occasion been discussed even in the distant past with quite disconcerting frankness. But Plato's Noble Lie and Machiavelli's good laws *and* good arms (an Industrial Relations Act *and* the Special Air Services) were formulae addressed to societies too different from at least the industrial societies of today to be very helpfully directive to a modern social theorist. The

[3] But cf. Blaise Pascal, *Pensées,* ed. L. Lafuma, Éditions du Seuil, Paris, 1962, 103, pp. 63–4: '+ Justice, force. Il est juste que ce qui est juste soit suivi; il est nécessaire que ce qui est le plus fort soit suivi. La justice sans la force est impuissante, la force sans la justice est tyrannique. La justice sans force est contredite, parce qu'il y a toujours des méchants. La force sans la justice est accusée. Il faut donc mettre ensemble la justice et la force, et pour cela faire que ce qui est juste soit fort ou que ce qui est fort soit juste. La justice est sujette à dispute. La force est très reconnaissable et sans dispute. Aussi on n'a pu donner la force à la justice, parce que la force a contredit la justice et a dit qu'elle était injuste, et a dit que c'était elle qui était juste. Et ainsi ne pouvant faire que ce qui est juste fût fort on a fait que ce qui est fort fût juste.' And see *Pensées* 60 (pp. 51–2), 61 (p. 53), 81 (p. 57), and 85 (pp. 58–9).

modern social theoriest who has made the most extensive and adventurous effort to think through this question is probably Jürgen Habermas.[4] It would be convenient to be able to present my own conception of this issue in his terms. But to do so would require a clearer understanding of his views than I can muster. Accordingly I pose the question instead in more homespun fashion for myself. It is not, it must be confessed, at all an easy question to pose without inadvertently begging it.

The key question is what one takes to be prior constraints on any valid social theory.[5] If, for example, one presumed, as methodological individualists in general would presume, that what must be taken theoretically as given in social theory is the consolidated existential substance (the experienced reality) of individual human lives, then society as such would be conceived as the pragmatic interrelations between such existences, externally, contingently, and causally. Social theory would be a purely causal mode of thought and the question of how false and how true politically relevant opinion

---

[4] See particularly Jürgen Habermas, *Toward a Rational Society*, translated J. J. Shapiro, Heinemann, London, 1971; *Knowledge and Human Interests*, trans. J. J. Shapiro, Heinemann, London, 1972; *Theory and Practice*, trans. J. Viertel, Heinemann, London, 1974; *Leigitimation Crisis*, trans. T. McCarthy, Heinemann, London, 1976. For historical and analytical introductions to his work see David Held, *Introduction to Critical Theory: Horkheimer to Habermas*, Hutchinson, London, 1980, and Thomas McCarthy, *The Critical Theory of Jürgen Habermas*, Hutchinson, London, 1978.

[5] Even amateur social theories are theories in the strong sense that they aim at explanation, prediction, or control. It is controversial whether there are ontological and moral constraints on the validity of social theories or whether the sole criterion of their validity is simply pragmatic. The view that human societies consist, *inter alia*, of all the human beings who constitute them, with all and only the beliefs and all and only the sentiments which these human beings happen to possess is not a social *theory*. It is simply a tautology. Because of the very evident pragmatic grounding of this tautology in the experience of all human beings, it is a constraint on any social theory that it preserve the full pragmatic force of this commonsense theory of social constitution (just as it is a constraint on any physical theory, however initially counterintuitive, that it preserve the full pragmatic force of the commonsense theory of material objects). The moral grounds for preserving this tautology, whether or not these are genuinely independent of the pragmatic grounds for doing so, are, as argued below, at least equally strong. (See John Dunn, *Political Obligation in its Historical Context*, Cambridge University Press, Cambridge, 1980, ch. 5.)

in modern societies has to be or could be would be merely a question in natural science. It would not in fact be a question to which we would ever in practice be likely to know an accurate answer. But there would be nothing intrinsically bemusing about what type of question it was. For a more robust type of wholist, by contrast, what is theoretically given would be located very differently. Substantive human reality would be society considered as a causal system and individual human lives would be just the medium in which its causal properties were implemented. And because reality in this conception lies elsewhere, and because human beings in general as self-conceived do not recognize it to do so, human lives as self-conceived would in essence be nothing more than tissues of more or less deeply felt illusion.[6]

There is some disagreement, naturally, within this view as to whether all forms of human society are necessarily constituted by deluded persons or whether there could, or even will, come to be in some settings a type of society which would be veridically transparent to its members and would have its causal properties veridically mirrored in the consciousness of these members and rationally endorsed (or even chosen) by them accordingly. This question also can be (and indeed at times has been) conceived as a question in natural science, the science of history. But it is worth noticing how hard it is even to formulate it without employing fairly blatantly ethical terms. Thus expressed, these two broad theoretical conceptions, the individualist and the wholist, are diametrically opposite. But they do have something massive in common. Each of them is in one respect deeply alienated, seeing a constitutive aspect of human existence resolutely from the outside and implicitly dehumanizing it by doing so. Methodological individualists see human sociality as an external and contingent attribute of human beings and wholists return the compliment by seeing human individuality

[6] For an interesting discussion of this question, particularly in relation to the work of Louis Althusser, see W. E. Connolly, *Appearance and Reality in Politics*, Cambridge University Press, Cambridge, 1981, esp. pp. 48–62.

as the fantasy of a creature constitutionally unable to apprehend its rigidly social location.

Now, each of these perspectives can and does disclose
important features of the human situation. And since each
can and at times does do so it is tempting in the first place
to opt for the anodyne resolution that they should be seen
not as antithetical but instead as complementary. This may
appear a characteristically commonsensical Anglo-Saxon
compromise, trading off a little mild intellectual scruffiness
for the advantages of a broad and undogmatic view. Its only
major inconvenience, perhaps, is that the two conceptions do
at first glance appear in fact to contradict one another — and
more importantly that this apparent contradiction does not,
on closer inspection, dissolve effortlessly, displaying the
irretrievable superficiality of the Anglo-Saxon understanding
and its brute insensibility to the truths of dialectics. The
reason this is so is extremely important. What lies behind
the explicit contradictoriness of theories, each of which is
founded upon what might well — and indeed should — seem
an indisputable attribute of human beings, is a fundamental
choice as to how to cast the theories in question. The motivation of this choice is a very obscure matter; but it is plain
that what it goes back to in the history of western epistemology
is the project, fathered particularly by Descartes, of seeking
to specify and comprehend what Bernard Williams has termed
an 'absolute conception of reality',[7] a conception of possible
objects of knowledge — the way the world and all that is in it
is — which is in no way relativized to the causal contingencies
of human experience. Whether or not this is a coherent
project in relation to any possible objects of knowledge —
mathematical, natural scientific, and so on — there are very
good reasons for not regarding it as coherent in relation to
many of the attributes of human beings.

Convinced methodological wholists and convinced methodological individualists both seek a fundamentally external

---

[7] Bernard Williams, *Descartes: The Project of Pure Enquiry*, Penguin Books,
Harmondsworth, 1978, esp. pp. 65-7, 245-9, 301-3.

view of the character of human beings in society, not simply as a pragmatic and temporary intellectual convenience (a choice which would be easy enough to defend) but rather as a presumptive precondition for attaining *knowledge* about human beings in society. On a wholist view individual self-understanding is simply a factual given and a genuinely epistemic status is only open to it insofar as it furnishes an accurate apprehension of an individual's objective social location. On an individualist view such self-understanding is also potentially a factual given. But it is a factual given precisely because, in some not very clearly demarcated respects, it is epistemically incorrigible in principle; and social causality is conveived as operating around its edges, externally to it and in response to its causal weight.

Now, human beings certainly do possess self-understandings. Human self-understanding is existentially actual. We all do have our own understandings of ourselves, however vague and prevaricatory and externally grotesque these may be. And the self-understanding of all human beings is deeply interwoven with their understanding of the social settings of their lives, with, that is to say, their amateur social theories. (Amateur social theory has always been, and will always remain, partly constitutive of human existence.) But although it is clearly appropriate, reckoning in the idiom of more or less brute fact, to take the self-understanding of human beings simply as existentially actual, it is also quite clearly inappropriate to confine oneself, in seeking to conceive such self-understanding, to an idiom of more or less brute fact. Human beings understand themselves the way they do and not differently. But they notoriously do not for the most part understand themselves especially well. (You may understand yourselves perfectly. But I must confess to finding much of myself a fairly impenetrable mystery.) And because all persons' self-understanding is predicated in part on their amateur social theory, on their conception of the meaning and pragmatic character of their more or less intimate or distant relations with many other human beings, this opacity in self-understanding is not just something crudely internal to their own minds or bodies, a matter of the deep penetralia of the unconscious or some theoretical surrogate for this. It is, in addition, a relatively simple function of the limits of their social vision, a direct mark of the limitations of their amateur social theories and the miscellaneous contingencies of their personal social experience.

Putting the point still more crassly, it is not possible satisfactorily to defend the view that the way to understand human beings is to think of them as creatures whose properties stop at the edges of their own bodies. For a long time this view drew powerful support from an interpretation of the epistemological implications of the scientific revolution. But today its popularity is plainly on the wane and it is hard to see it ever being resuscitated simply on grounds of intellectual plausibility. The main individualist alternative to this view, that we should think of human beings above all as creatures whose properties stop at the edges of their own minds, also draws support from an interpretation of the epistemological implications of the scientific revolution; and, unlike the view that human beings are in essence simply physically bounded creatures set within a physical universe, this more mentalist conception is today very much on the advance — and indeed wreaking some little havoc even within the philosophy of natural science itself. But what is important for my purpose is not the adequacy or precision of such a mentalist conception of the human individual, still less its current intellectual popularity, but rather the fact that even if it is in principle adequate and can in practice be made precise (two very optimistic assumptions), the edges of our minds today stretch so far spatially and temporally that they take in, however fecklessly and ineffectively, a large portion of human experience as a whole. This is not, of course, to say that each of us in any sense understands human experience as a whole, a notion which, if intelligible at all lacks any shred of plausibility. What it means, rather, is that our self-understanding today is predicated upon, incorporates references to, forms of human social interaction which stretch across the entire globe and forms of temporal awareness which take in many past generations and at least some hoped for (or feared) future generations. Our self-understanding today has this bewildering (and indeed bemused) spatial and chronological extension, of course, because of the extent to which, and the pace at which, the world as a whole has changed — and changed over the last century in particular. It is important, too, to realize that this overextended consciousness which today we have no choice but to seek somehow to incorporate into our conceptions of ourselves is not a prerogative of, or a burden confined to, intellectuals at major universities, but, rather, a fate incumbent on the vast majority of human beings now alive.

Even in the furthest recesses of the Ghanaian rain forest today illiterate cocoa labourers know quite well that the social causality within which their lives must be lived out is bounded not by the hamlets in which they reside nor by the forest itself nor even the territorial borders of the state of Ghana (which many of them have crossed in order to secure a cash income at all) but rather by the world cocoa market and the consuming tastes and habits and capacities of the populations of very distant and alien countries. No doubt most of them understand this vast and intricate causal field very poorly indeed. (Don't we all in our own case? You and I, Sir Keith Joseph, and Mr Tony Benn? International economic exchange is the major witchcraft zone of modern experience.) But what matters is that even in the depths of rural Africa modern selves, the selves of today, *must* incorporate into their amateur social theories, in however gingerly a fashion, a conception of how global social and economic causality bears upon their own lives.[8]

All of this, to be sure, is only a fact about modern history. In itself it may cast a little doubt on the practical wisdom of picking individuals as the explanatory units of social theory merely on grounds of their palpable determinacy; but it

---

[8] For an attempt to picture this transformation over the last century see J. Dunn and A. F. Robertson, *Dependence and Opportunity: Political Change in Ahafo*, Cambridge University Press, Cambridge, 1973.

implies nothing at all about the fundamental terms in which human beings have good reason to understand their societies and their selves. Once it is recognized as a fact about modern history, however, and once we have drawn the implications which it presents for the forlornness of the major extant traditions of social understanding, we can perhaps reconsider the possibilities for social understanding in a somewhat humbler and more patient mood. And perhaps, if we are lucky and persevere, we may even eventually engender a form of social understanding which is less heavily toxic in its impact.

I have tried in outline to trace the process of intoxication to two very different sorts of factor, one clearly an aspect of the history of thought and the other more crudely an aspect of the history of society. The peculiar and menacing arrogance of modern social theory is a product, I have tried to suggest, of grandiose epistemological pretension and of the place of university teachers and secular intellecutals more generally in the modern division of labour.[9] Neither of these factors is arbitrary or irrational. But it is a simple and sober judgement to make by now that, at least in the field of social theory, each has greatly overreached itself. Social authority in the modern world (at least when explicitly taking the initiative) takes its stand on the presumed cognitive prowess of intellectuals: Lenin or Hayek, Milton Friedman or Stuart Holland. It does so insofar as it both needs, and recognizes the need, to understand what it is doing. And intellectuals duly respond by casting their beliefs as to what is desirable

---

[9] These social roles do not, of course, compel their occupants to adopt views of such arrogance. But they do foster an arrogant culture. Just as philosophers, in Richard Rorty's jibe, see themselves as 'knowing something about knowing which nobody else knows so well' (Rorty, *Philosophy and the Mirror of Nature*, Basil Blackwell, Oxford, 1980, p. 392), so social theorists and social scientists picture themselves as knowing something about knowing about society which no one else knows so well. Perhaps in many instances, indeed, they do; but like philosophers, they find it very hard to specify just what. The presumed possession of this happy epistemic knack is perhaps only an occupational disease. But it is apparent how hard it is to distinguish the occupational disease from the occupation.

or possible in a cognitively professional idiom, modelled as best they can on the sciences of nature.

III

It would not, needless to say, be a better idea to seek instead to found social authority today on explicit whim, or to replace the advice of secular intellectuals in its entirety by that of astrologers. But it certainly would be an improvement, a lengthy stride towards detoxification, for professional social theorists to learn to express their guesses as to what is desirable or possible in a more intellectually democratic and a less cognitively pretentious idiom.[10] In social theory at least, the charms of the absolute standpoint have been strictly those of an illicit title to authority. The eyes of God offer a viewpoint which may well have been a necessary condition for the very idea of the absolute conception. But human beings cannot see themselves through the eyes of God and not even social theorists are well advised to claim to see their fellows from this vantage point. Pragmatically considered through the eyes of God, the space of human existence in its entirety, perhaps, should be taken simply as a homogeneous causal field and one which is not ontologically broken up into discrete units at the level of individual or society. But we, as human beings, certainly have no idea of how to conceive it as such; and we have at present no good reason to believe that to succeed in conceiving it accurately as such lies within the scope of human cognitive powers at all. To design social theories to meet such an epistemological standard is in itself merely to fantasize an imaginary cognitive potency for ourselves. But by doing so we risk worse consequences than that of simply making fools of ourselves. In particular we risk coming to believe in our possession of such powers and coming to feel a resulting entitlement to thrust aside the amateur social theories of

---

[10] Cf. Dunn, *Political Obligation in its Historical Context*, Ch. 10, esp. pp. 284-5, 287-9; Rorty, *Philosophy and the Mirror of Nature*, pp. 376-7. And cf. Williams, *Descartes*, pp. 162, 188, 195, 210, 268-9, etc.

others (and the professional social theories of our rivals) in order to implement the epistemic insight which we ourselves supposedly possess.

It could not, of course, generally make sense to will our society to be organized and governed on the basis of ends which we do not favour and causal beliefs which we suppose to be false. Even the most professional social theorist is an amateur social theorist under the skin. In the first instance, at least, one amateur social theorist, one vote. And, within the relevant social circles, to will defeat for the presumptively false professional social theories of our rivals is simply a criterion of sincerity for our believing our own social beliefs. But the relation between our own professional social theories and the amateur social theories of others is a good deal more delicate. Any purely causal social theory will have to treat much amateur social theory in a decisively external way. But no human being (professional social theorist or not) has any right because she or he regards the beliefs of other human beings from a causal point of view simply and exclusively as matters of fact and from an epistemic point of view simply as false beliefs, to treat them from a practical point of view as politically illicit and lacking in weight.[11] Or if, for harshly Machiavellian reasons, any human being does have such a right, she or he possesses it not *qua* professional social theorist but *qua* amateur social theorist, doing the best she or he can in the face of the odious importunities of politics. *Salus populi suprema lex.* The worst temptation in social theory, greatly exacerbated by academic professionalization and by the increasing political prominence of social theory in the practical reproduction of modern societies, is to treat a social theory as an external and technical device for augmenting the political entitlements of the self or depleting the real identities of other human beings.[12] To resist this temptation at all effectively it is above all else necessary to try to form a conception of the

---

[11] See Dunn, *Political Obligation*, Ch. 5.
[12] See Connolly, *Appearance and Reality*.

relations between identity and social causality in human beings which can be trusted to remain fairly stable when applied both to other human beings and to one's own self.

In relation to ourselves we certainly cannot afford to regard social theories merely as wholly external intellectual devices because we in fact conceive ourselves and fashion ourselves partly through the social theories which we form. (On a behaviourist vision of human nature this might seem a somewhat murky thought; but on a more mentalist view it could scarcely be more blatantly obvious.) No more can we afford to conceive human beings (including ourselves) as entities epistemically external to the vagaries of their actual consciousness. We cannot afford to do so, not for reasons of compulsive narcissism – because to do so would inflict such painful wounds on our self-esteem – but because to do so would jam human practical reason across the board. Human beings could not act at all without possessing some conception of themselves as agents; and they could not have good reason to act without possessing conceptions of themselves as agents which they have good reason to possess.

Social theory is a branch of human practical reason and its obsession with epistemology over the last three centuries, however historically intelligible, has made it serve the requirements of human practice very poorly indeed. Only if we learn to expect very much less of it do we stand a reasonable chance of refashioning it so that it can serve us rather better. At least one constraint on any humanly acceptable social theory is that it should not impair the adequacy of the conceptions which its exponents hold either of themselves, or of other human beings, *as* human agents. It should, that is to say, at the very least exclude unjustified contempt or sentimentality *throughout* its conceptions of what human action is like and of why this occurs as it does. The importance of closing the gap between the understanding of the self and the understanding of others is obvious enough here; and it requires no great feat of the imagination to grasp the potential political implications of the persistence of such gaps. (To take only a single example: one can be confident that human beings cannot in practice dispense with a conception of themselves as agents. If therefore they adopt a social theory (such as that of Professor Althusser) which repudiates such a conception in its entirety, what they will in fact be left with is a wholly external (and implicitly contemptuous) vision of the agency of most human beings and a vision of themselves and perhaps of their more intimate political associates in which this external and contemptuous vision is compulsively suspended, to be replaced by one which could scarcely avoid sentimentality and self-deception.)

Within political practice, very broadly considered, the role of social theory over the last two centuries has been at least as much one of assigning blame as it has one of guiding political action. In itself a purely causal and instrumental social theory (if such could be discovered) might at first glance seem well enough suited to the guidance of political action. But one need not be surprised that such social theories mutate fairly drastically under the pressure of the demand to assign blame for the more unacceptable aspects of history as directly encountered. Powerful theories of whom to blame soon construct strong categorical distinctions between trustworthy selves and untrustworthy others; and by doing so they augment their sparely causal contentions with expressively more implicating facilities for human identification. The initial pressures towards this mutation may be instrumental and clear-headed enough. But their consequences rapidly cease to be so. Causal understanding and corporate political self-righteousness become steadily harder to disentangle. What begins as a means for understanding the world and for recruiting followers to share this understanding becomes in due course a hysterical compound of projective suspicion and analytical confusion, in which political entrepreneurs are every bit as deeply mired as those whom they have previously persuaded and in which political self-righteousness drastically impairs causal understanding.

IV

Here, plainly enough, we come to the politics of today, albeit very briefly and in no very constructive spirit. (If I wanted to get anywhere at all to which I could genuinely wish to make my way, I would certainly not start from here and now.) There is little dispute today that this country is in a very bad way and rather little expectation that it will avoid soon being in a decidedly worse one. The causes of a great deal of its predicament are by now rather well understood. Many of them, indeed, are frequently mentioned in public discussion. A few even appear in the political debate between the present government of the country under Mrs Thatcher and its only politically viable alternative — those bits which fit fairly effortlessly into a justification of the demented policies of this government and, by contrast, those bits which can serve to highlight some of the more reckless acts of vandalism which it has perpetrated. The disparity between the extent of available understanding and the paucity of this understanding which is politically espoused and deployed is extremely striking — and all the more striking because the intensity of the social crisis which we face, both immediately and in the longer term, is no longer seriously denied by any political actors. No doubt the degree of omission and caricature in these political interpretations of what is to be done here and now is a fairly direct product of the intensity of the crisis. In routine political conditions a largely undirective official social theory or one which simply endorses a stolid persistence in current practices is easy to accept. In conditions of crisis the view that something in particular needs to be done is harder to resist. The two candidates for official social theories at present (the political visions of the dominant groups in the two major parties) accordingly each prescribe very drastic actions. Each promises in effect to the amateur social theorists who make up the electorate (and indeed the labour force) that if it is permitted to act as it wishes (to make its social theory the official social theory and implement this through the machinery of government) and if the citizens behave as it enjoins them, the predicament of the country can be successfully eluded. In each case the social theories offered, insofar as they are determinate, clearly owe much to the past labours of professional social theorists: in the case of the

Thatcher government, whose policies are in a number of ways unusually determinate (if not particularly coherent), to some variety or other of monetarist economics; in the case of the Labour Party at present to a more nebulous range of intellectual sources. In neither case do the social theories as publicly proffered posses a shred of intellectual plausibility as a means for dealing with the present difficulties of the country, let alone with the problems which it is likely to face in the future.

It is easiest to discuss the main defects of the present government's vision since it is so hard to get clear answers about anything from the present leadership of the Labour Party and so difficult to be confident where precisely the leadership of the Labour Party should now be deemed to reside. (All that can perhaps be said on the last score is that it is extremely hard to believe that a combination of high tariff barriers, extensive nationalization, and freezing existing labour allocation in an economy administered by the man who gave us our impressive fleet of Concordes is a promising guarantee in even the medium-term future for the existing living standards of the British working class.) About the Thatcher government, however, it is easy to say all too much. I will confine myself to making two points. The first is about the incoherence of Mrs Thatcher's economic views as a *political* doctrine. The view that high interest rates and a commitment to lowering public expenditure would in due course force down the level of wage settlements because of the huge rise in unemployment it would threaten is quite incompatible with the general theory of economic motivation to which Mrs Thatcher is committed. On this theory, for the level of unemployment to act as a threat in any particular wage negotiation it must be the unemployment of the workers negotiating which is directly in question. As a political threat, therefore, the menace of unemployment can only hold down wage levels rapidly and generally if the working class shows an extraordinarily high level of class solidarity — a level of class solidarity which is utterly incompatible with the theory of economic motivation on which all Mrs Thatcher's policies

rest! It is, of course, true that to obliterate the British economy will have a substantial effect on wage settlements,
along with the rest of our lives. But it is important to try to
remember that the Thatcher government's policies began as
a comparatively rational (if misguided) attempt to change
British economic activities and not simply to end them.

The second point is a little less crude. It involves in the
first instance inquiring innocently why it was that Mrs
Thatcher (and Sir Keith Joseph)[13] ever expected the organized
British working class to exhibit such extraordinary class
solidarity. The explanation, it seems likely, lies in the extreme
simplicity of Mrs Thatcher's social vision and the extent to
which she did genuinely perceive the organized working class
as a single bloc adversary, in the short term at least a unitary
political enemy. Within the legally somewhat underpoliced
zone of civil society, the zone of family life and of production
but also of flying pickets, the past votes and the industrial
power of the working class had established conditions in
which profitable production was impossible. But the past
votes of the working class were of course civic actions, actions
directed at the order of public law and the state; and within
the order of public law and the state the British working class
since 1945 has attained a reasonably high and stable level
of class solidarity. An economic policy which could lower
real wages and a political thrust which could transfer civic
dutifulness and a long-term sense of collective interest from
an undeserving social democratic state to a very abstract
conception of rational conduct in the labour market, taken
together, might do wonders. Class solidarity, at least amongst
the working class, could be evicted from politics where it
could only do harm and transposed deftly into private
economic conduct where it would take as its goal the long-
term public good of restoring profitability throughout the
economy, instead of the short-term private good of maximizing
individual wages. This plainly is the politics of fantasy — and

[13] Some more direct light is cast on Sir Keith Joseph's views in Sir Keith Joseph
and Jonathan Sumption, *Equality*, John Murray, London, 1979.

at least as much so as the celebrated views of Mr Benn.

Few, perhaps, would accept the justice of this description of the premises of Mrs Thatcher's policies. But however ungenerous it may seem as a description of the Prime Minister's better intentions, it does, I trust, bring out the potential perils of taking stray bits and pieces of professional social theory and pretending, at first to others but then no doubt rapidly to oneself, that these stray bits and pieces do genuinely articulate the causal dynamics of the extraordinarily complicated space which extends from all of us as amateur social theorists to the structures of the nation state of which we are citizens and the international economy on which we haplessly depend for our livelihoods. What requires emphasis is not the peril of uniting such shreds and tatters of causal understanding with power, a union to which there may in any case be no real causal alternative, since full causal understanding of society may be naturally impossible for human beings and indeed the very idea of such understanding for human beings may simply be internally incoherent. What must be stressed, rather, is the practical political importance of seeing whatever shreds and tatters of causal understanding a government can muster in direct relation to the amateur social theories of ordinary citizens. To see what, if anything, is to be done here and now is to see on what terms and for what purposes we could rationally co-operate with each other. To implement such a vision in practice (supposing it to be accurate) would be above all to discover how in practice we could learn to trust one another to co-operate. Because what is in question is what is to be done, the historical constraints on the second (roughly, the class history of the country) are also constraints on the first. Perhaps in fact over any lengthy period of time there simply *are* no such terms, or no such terms which include even the great bulk of the population – in which case the actual future of this country is likely to be grim. A hazy sense of these points is in fact rather prominent in the public pronouncements of Mr Benn, though it does not appear to be accompanied there by any serious respect for social causality at all. Neither Mrs Thatcher's sincere refusal to listen to the people, nor Mr. Benn's insincere eagerness to be instructed by them, exemplifies a very satisfactory conception of the role of social theory in political action; but each does express a real feeling for the major blemishes of the other.

V

I have tramped a number of times round a rather small cage and I certainly cannot promise in conclusion any neat way out through the bars. But what I would like to do in conclusion is to consider once again very briefly the relations between amateur, professional, and official social theories and political action and to do so in the context of the very old-fashioned questions in political theory and ethics of how we have good reason to perceive ourselves and how it makes sense for us to live our lives. We may take these last two questions very much in the mood of the Platonic dialogues, particularly the *Gorgias* and the *Republic* — not, of course, because the theoretical answers which Plato offers to them in those dialogues are either clear or convincing — but because the questions themselves have never been expressed with more urgency. If the way in which we have good reason to live our own lives is, as Socrates argues against Callicles, not as more or less discreet brigands[14] but rather with the principled trustworthiness which is a precondition for friendship and community, then the same should hold good of other human beings too. And if this is not the way in which we have good reason to live our lives, then we cannot reasonably expect others to presume it appropriate for themselves either. Civic dutifulness and a genuinely egoistic maximization can be fitted together, both in theory and in practice, only by a measure of intellectual inattention.[15] And this is true over time either way — whether we adopt sentimental or moralistic views of ourselves as moral agents and cynical views of others, or espouse our own cynicism self-referentially

---

[14] Cf. Plato, *Gorgias*, 507e (trans. and ed. Terence Irwin, Clarendon Press, Oxford, 1979, p. 86) and *passim*.
[15] This is a central issue in Connolly, *Appearance and Reality*, esp. pp. 125, 127.

and yet expect that others will pay the costs of the moral expectations which we solemnly proffer to them. On the whole professional social theories take the first form, while official social theories (since they have to solicit the co-operation of subjects or citizens and since governing is so troublesome) more frequently take the second. Professional social theories are very much theories cast in the third person. (You and I, we're good friends — and perhaps even good citizens if we choose to be so — but they — they're egoists to a person: just in social life for what they can get out of it.)

If we return once more to the political predicament of this country and ask again for what purposes we could rationally co-operate with one another — and trust each other to do so in actuality — the dangers of this split in our human vision become very apparent. Taking the class history of the country and the present organization of production within it as given (as for the moment we must) it is not very plausible that there are many purposes for which everyone does have good reason to co-operate with one another, and the possibilities of rational trust extending all the way across this space seem slimmer still. The intrinsic difficulties of making capitalist production as such an object of allegiance are immense and probably insuperable. If we consider in the first instance for what purposes we might rationally co-operate with one another as a question of what it makes sense for us politically to do, it is apparent that any third-person theory is ill-shaped to take the choice for us. Third-person theories can certainly draw our attention to important causal considerations — in particular to impossibilities and probabilities. They can tell us (or at least try to tell us) what might work and what certainly cannot be brought about. But there is not, and could not be, anything else to tell us politically what to *do* but the amateur social theories which we possess as ordinary agents.[16] Collective political and social life (even the repro-duction of the capitalist mode of production) is not something

---

[16] I have tried to bring out the fundamental significance of this point in political theory in *Political Obligation*, Ch. 10.

which can be done for us by a government or a party. It is the *living* of *our* lives. The living of our lives is something for which professional social theory has (and perhaps can in principle have) very little respect. And at least when politically applied from above (when espoused as official social theory) what professional social theory does is to expropriate the existential reality of individual lives. As official social theories, as doctrines of state, all professional social theories are wildly undemocratic — whatever formal role they may allot to the term democracy in their public self-justifications. Democracy is simply the political form of fraternity, the recognition that the species to which we belong is a species comprised of amateur social theorists and that rational co-operation and trust for its members must always depend in large part upon their actual beliefs and sentiments. Fraternity is a very varying historical possibility. But it can only *start* where professional social theory leaves off.

It is not that there is something misguided or vicious in the attempt to understand social causality — or indeed that there is any alternative to making such an attempt as best we all can. Still less is it the case that social causality is potentially to be eluded or is somehow unreal. The view that there is no such thing as social causality could not in fact form part of the amateur social theory of anyone at all. One could not hold it and live a human life. The fantasy of a wholly causally transparent social world, a world so clear that life within it would have neither need nor opportunity for trust, is simply a political sedative. Men, as John Locke said, *live* upon trust.[17] They behave as they behave and act as they act on grounds which are necessarily for the most part insufficient. Rational political co-operation amongst them requires a great deal of courage and good will. But they cannot be offered any rationally superior alternative to political co-operation. Political authority today in most countries has for the most part given up the sacred. But to replace this, where it retains

---

[17] *The Correspondence of John Locke*, ed, E. S. de Beer, Clarendon Press, Oxford, Vol. 1, 1976, p. 123.

any ideological pretensions at all, it has clutched to itself instead a style of social theory which in epistemic terms is merely a pale shadow of the sacred. It has done so, in effect, in order to retain the key prerogative of authority, the entitlement to speak *de haut en bas*. It is certainly a government's business to attempt as best it can to understand social causality. But it is equally certain that any government's success in understanding social causality will be extremely limited. All a public authority can offer its citizens, at the level of intention, in this respect is the sincere attempt to understand social causality and the frank avowal of such understanding as it does hold. Rational co-operation between citizens and government will necessarily depend upon how cogently this understanding meshes with and modifies the amateur social theories of the citizens at large. Political authority can certainly be heuristic. But it cannot appropriately and in general be didactic, because there is nothing determinate for it to teach. Collective social and political life is not a classroom, with masters (or mistresses) and pupils. In the end there is simply *us*, trying to decide as best we can and on the basis of necessarily limited powers, what we have good reason to do. It is this more modest and democratic vision of the nature of modern political authority which we would have learnt by now if we truly understood the history of social theory. It is this to which we need individually and collectively to discipline our imaginations. And unless we in this country do manage collectively to learn it and learn it very fast indeed, our political and social future here is likely to be excessively ugly.

The Collapse of Consensus:
Ideology in British Politics

David Marquand

It is clear to the serious student of modern politics that a mixed economy is what most people of the West would prefer. It is neither prudent, nor does it accord with our conception of the future, that all forms of private property should live under perpetual threat. In almost all types of human society, different forms of property have lived side by side without fatal consequences either for society or for one of them.

Aneurin Bevan, In Place of Fear, 1951.

The public are bored with the old terms of Private Enterprise and Socialism, Protection and Free Trade, but they are interested in any movement that will give them what they believe can be done with the vast resources — technical and scientific — that we have at our command. Sooner or later either this democracy will die or we shall be able to lead this country steadily, not by revolutionary, but still by progressive and ambitious, means into the realization of the new society which is open for us if we have the courage to seize the opportunities that are ready for us today.

Harold Macmillan, Speech in House of Commons, 1938.

Post-war experience, especially in the United States, supports the view that measures to maintain or restore demand are both practicable and broadly effective. The modern economist's income analysis actually seems to work. It is not always possible to prevent the intitial occurrence of minor recessions, but the significant fact is that these can now be quickly corrected and in a broadly predictable manner. The prosperity of the United States since 1945, maintained in the face of unceasing prophecies of the imminence of depression, is proof at once of the strong autonomous pressures towards full employment and of the power of the modern state to reinforce these as necessary.

Anthony Crosland, The Future of Socialism, 1956.

I

It is a truism that British politics became more ideological
and less consensual in the seventies than they had been in the
fifties. It is not so clear why, or what consequences follow.
For the consensus of the fifties went wider than is sometimes
appreciated. As my three quotations all imply in different
ways, central to it was a certain conception of the proper role
of government — a conception which was both positive and
negative. Positively, it was accepted that the state had three —
and later four — crucial functions, all of which had been
highly controversial twenty years before. In the first place, it
had a duty to provide certain welfare services to all its
citizens — notably education, medical care, and insurance
against unemployment. Secondly, it was taken for granted
that the state had a duty to ensure that no citizen fell below
a certain minimum standard of material provision; and that
that minimum should rise as society as a whole became more
prosperous. Thirdly, it was assumed that the state had a duty
to maintain full employment; and it was also assumed that
the Keynesian revolution in enconomics had given it the
means with which to do this. Economists, Anthony Crosland
argued in *The Future of Socialism*, now knew how to prevent
unemployment; the politicians' job was to put the economists'
advice into practice. Fourthly, it was also beginning to be
accepted, at any rate by the late fifties, that, in addition to
maintaining full employment, the state had a duty to speed
up the rate of economic growth. Opinion formers in this
country had begun to realize that Britain's competitors on
the mainland of Europe enjoyed faster rates of growth than
she did; though no-one knew how this evil could be cured,
few doubted that it was a task for the government, which
could not be left to market forces alone.

The negative part of the consensus was hinted at in my
quotation from Aneurin Bevan. It was taken for granted that
Britain had, and for the foreseeable future would continue
to have, a mixed economy, in which public and private
property existed side by side. In practice, moreover it was
also taken for granted (though not, it must be admitted, by

Bevan and his followers) that the boundary between the
public and private sectors should be that which had been
arrived at rather accidentally by the post-war Attlee government. To be sure, the boundary owed little to any obvious
logic. Over the years, the Labour Party had assembled a
number of commitments to nationalize particular industries.
These commitments owed much less to any general principle
than to the attitudes of the trade unions most directly
concerned. The miners were passionately anxious that the
coal mines should be nationalized, so the commitment to
nationalize the mines was high on Labour's list of priorities.
Chemical workers exhibited no such passion, so a commitment to the nationalization of the chemical industry made
only a brief and slightly shamefaced appearance. Despite
its illogical origins, however, the boundary between the
public and private sectors which had been reached between
1945 and 1950 was effectively sacrosanct in the fifties. The
Labour Party was committed to further public ownership
in principle, but had little stomach for it in practice. The
Conservative Party had long since been converted to the
heresies with which the young Harold Macmillan had shocked
it in the thirties, and although it opposed further nationalization it made only a fitful and half-hearted attempt to undo
the nationalization which had already occurred. There were
a few boundary disputes, notably in the steel industry, but
they were not very serious and were not fought with much
zeal.

    This consensus about the role of government as such went
hand in hand with a consensus about the nature and content
of the party struggle which determined the composition of
particular governments at particular times. It was accepted
that power should be monopolized by one of two great
parties, in constant competition for governmental office; that
victory in that competition would go to the party which
convinced the electorate that it was more likely to manage
the economy successfully; and that, in terms of outputs, at
any rate, the chief – perhaps the only – difference between
the parties was that they would distribute the fruits of

successful economic management in different ways. The Labour Party was known to stand, broadly speaking, for the working class, and the Conservative Party, speaking even more broadly, for other interests, less precisely defined. It was assumed that Labour governments would be more tender towards the working class than Conservative governments, and that Conservative governments would be more tender towards non-working-class interests than Labour governments. It was also assumed, however, that the conflicts between the working class and the class or classes whose interests were served by the Conservative Party were, in the jargon of social science, 'non-zero-sum' conflicts: that it was possible to arrive at compromises between them, from which both would gain: and that both therefore had a stake in the survival of the system, in a sense which the losers had not had (or had not thought they had) between the wars.

Consensus about the content of politics was buttressed by consensus about the rules of the political game. Britain had seen no serious constitutional crisis since before the First World War. In the early thirties, it is true, the left wing of the Labour Party, led by Sir Stafford Cripps, had suggested that an incoming Labour Government might meet so much resistance from organized capital that it would have to take emergency powers and suspend the normal operation of the parliamentary system. It turned out that the post-war Labour governement was able to carry through a vast programme of what then seemed o be very far-reaching reforms, while making no significant changes in the constitution which it had inherited. Four aspects of that constitution deserve particular attention. In the first place, parliament was absolutely sovereign, in a sense true of no other parliaments. Not only were there no judicial checks on its freedom of action, but in principle there could be none. Secondly, ministers and ministers alone were responsible for the government's policy decisions; their officials were assumed to be political castrates, whose sole functions were to advise ministers before decisions were taken and to adminster the results. Thirdly, the United Kingdom apart from Northern Ireland – whose constitutional status could have been changed at any moment at the sole wish of the Westminster Parliament — was both a resolutely unitary state in law, and an astonishingly centralized state in practice. There were no provincial or regional legislatures; though Scotland and Wales were culturally distinct from England, they had no vestige of political autonomy; even in great cities like Manchester, Birmingham, and Liverpool, the local authorities were, in practice, little more than agents of Whitehall. Fourthly, this huge agglomeration of power was supposed to be made accountable to the public through a continuous adversarial struggle between the two great parties on the floor of the House of Commons — whose function as a forum for combat between two teams of political gladiators, competing for votes at the next election, had thus become far more important in practice than its older functions as a legislature and as a watchdog over the executive.

That was the formal constitution, sometimes referred to as the 'Westminster Model': the constitution of the legal textbooks and of politicians' speeches. Under its skin, however, a new set of arrangements had grown up, the assumptions of which often ran counter to those of the formal constitution. Since the First World War, as Keith Middlemas has shown,[1] a loose, informal, fluctuating system of *de facto* power-sharing had come into existence between the state, as the representative of the general interest, and the great producer groups, particularly the trade unions and the representatives of the organized employers. This system of 'corporate bias', as Middlemas called it, was not based on any formal ideology, though it was sometimes justified in ideological terms. Governments slipped into it because they discoverd in practice that they could not manage the country's business in any other way. They needed the co-operation of organized labour, in the first place to win the First World War. They still needed it even in the twenties and thirties; they needed it far more in the forties and fifties. And although the cor-

[1] Keith Middlemas, *Politics in Industrial Society: the Experience of the British System since 1911*, André Deutsch, London, 1979.

poratist system was never underpinned by any formal ideology, a corporatist quasi-ideology gradually won acceptance after the system had developed. By the fifties, it had come to be believed that, in some sense or other, decisions taken by government were not fully legitimate unless they had emerged from corporatist consultations between the state and the organized groups concerned. Even Aneurin Bevan, an instinctive anti-corporatist, had spent long hours in negotiations with the doctors before setting up the National Health Service. One reason, of course, was that the Health Service would not have worked if the doctors had refused to work it. I doubt, however, if that were the only reason. It was also felt that it would be wrong for a government to introduce radical changes without consulting the interest groups most obviously involved. Yet that assumption, though not absolutely contrary to, was clearly at variance with, the fundamental assumption of the formal constitution – the assumption that parliament was and ought to have been absolutely sovereign because it, and it alone, properly represented the people.

Such, then, was the consensus of the fifties. It was not a universal consensus. On the left wing of the Labour Party there was still a muddled, though strongly felt, attachment to the notion summed up in Clause Four of the Labour Party constitution: the notion that social justice is impossible so long as the means of production are privately owned. That attachment was not felt strongly enough, or by enough people, to make much difference to the proposals actually put forward by the Labour Party in general elections. It was, however, held strongly enough to prevent Hugh Gaitskell from revising the party's old 1918 constitution, and to ensure that he and his colleagues remained formally committed to social ownership for its own sake. Moreover, there was still, both on the left wing of the Labour Party and on the left wing of the trade-union movement, a residual attachment to a different conception of the relationship between class and politics. In and near to the Labour Movement there were still people who saw the class struggle in the old 'zero-sum' terms of the inter-war years: who believed, that is, that there was an irreconcilable conflict of interest between 'haves' and 'have-nots'. They were too weak to prevail in the internal party struggles of the fifties, but they were there; and when a change took place in the leadership of the Transport and General Workers Union — the most powerful union in the Labour Party then, as it had been since the 1930s — it became obvious that they were still quite numerous. By the same token, there were still people in the Conservative Party who did not accept, or who did not wholly accept, the Keynesian revolution and who rejected the 'middle way' of Harold Macmillan, as enunciated in my quotation from him. That became clear in the famous 'little local difficulty' when Thorneycroft, Powell, and Nigel Birch resigned from the Treasury in 1957 in protest against increases in public expenditure.

But although there were dissenters from the consensus in both parties they were effectively 'marginalized'. They hovered menacingly in the wings, mouthing imprecations at the players in the lead roles, but they were rarely allowed on the stage. It would be wrong to suggest that they had no influence at all: the fact that they existed was a significant constraint on their leaders' freedom of action. Had the Labour Party contained no believers in public ownership for its own sake, Gaitskell and his allies might have turned it into a British S.P.D. Had the Conservative Party consisted exclusively of Macmillanite adherents of the 'middle way', it might have become a British C.D.U. In either event, the last twenty-five years of British history would certainly have been very different, and would probably have been much happier. But the dissenters exercised negative power, not positive. They could prevent others from doing things. They could not do things themselves.

II

That consensus has clearly broken down. In both the big parties, the marginalized dissenters of the fifties have re-turned from the wings to the centre of the stage. The present

government has abandoned the commitment to full employment, the single most important element in the old consensus,
and has come close to abandoning the commitment to maintain minimum levels of material provision. The Labour
opposition has virtually abandoned the commitment to the
mixed economy, and although it promises to return to full
employment, it no longer accepts the Keynesian orthodoxy
of a generation ago. Meanwhile, the 'Westminster Model' is
in disarray. Parliamentary sovereignty has been undermined
by the introduction of the popular referendum, and circumscribed by entry into a supranational European Community
whose laws take precedence over British Law, and whose
courts can set aside the judgements of Brtish courts. Though
the old doctrine of ministerial responsibility has not been
formally abandoned, the assumptions on which it was based,
in particular the assumptions of civil-service neutrality, have
become more and more implausible. Though the state is still
monopolized by the same two parties which monopolized it
in the forties and fifties, their hold on the electorate has
weakened to such a point that, according to the opinion polls,
a general election held at almost any time in the summer of
1981 would have returned a Liberal–Social Democrat alliance
with a majority in the House of Commons.

Clearly, something very odd has happened to British politics
in the last thirty years. What is it and what are its consequences
likely to be? The first and most obvious reason why the
consensus of the fifties has collapsed is that the economic
assumptions which underlay it have broken down. Both the
moderate, later self-consciously 'revisionist', Croslandite
Right of the Labour Party and the Macmillanite 'middle way'
Left of the Conservative Party based their claims to power
on their capacity to provide economic growth and full
employment, and to combine them with reasonable price
stability. Both assumed that they would be able to do this
because the world economy would continue to exhibit the
features described by Anthony Crosland in my quotation
from *The Future of Socialism*. The arrival of a worldwide
economic crisis in the early seventies, together with a worsening

economic performance on the part of this country in particular, cast doubt on all those claims, for the fundamental Keynesian assumption that it was possible to find a middle way between traditional capitalism and traditional socialism was now called into question.

At the same time, economic failure led to growing public dissatisfaction with the political system. The membership of both major parties of this country fell significantly. Though this cannot be proved, it is reasonable to assume that the people who have died or who have left the main parties were closer to the old consensus of the fifties than are the people who have remained in or have joined them. As a result, the membership of both parties is closer in attitude and spirit to the dissenters of the fifties than was the case a generation ago. This has narrowed the manoeuvring room of the party leaders, whose positions ultimately depend upon the support of their rank and file. Public dissatisfaction with the parties, moreover, has had a rather paradoxical effect on the parties' attitude to the primordial task of vote-gathering. They have become increasingly anxious to make sure of the solid reliable support of their own 'constituencies' before making forays out into the no-man's-land where other supporters might be found. In both parties, the result is a lurch back to the old inter-war attitudes to social class and to the relationship between social classes. The lurch is most obvious in the Labour Party. The harder it has found it to win over new supporters, the more determined it has been to hang on to its old ones; the more it has had to hang on to its old supporters, the more loudly it has had to sing the old songs which its old supporters want to hear. And the old songs are, so to speak, 'zero-sum' songs: songs which assume an irreconcilable clash of class interests and which reject the gentler, softer, and more fuzzy conceptions of a generation ago. The Conservatives' equivalent lurch is less dramatic, but no-one who has heard Mrs Thatcher appeal to the anti-union prejudices of a Conservative Party conference can doubt that it has taken place.

Public dissatisfaction with the system has had a further

consequence as well. The more disaffected the public becomes, the harder it is for governments to win the public's support, without which no economic policies will work. Failure breeds cynicism, and cynicism, failure. Because the public is cynical, the politicians have to promise more and more — or, at any rate, think they have to promise more and more — to win public support. The more they promise, the more promise diverges from performance; and the more cynical the public becomes.

The effects of economic failure have been reinforced by a complex change in the position of the trade unions. The notion of free collective bargaining has always been central to the unions' conception of themselves and of their role — central, that is to say, to the justification which union leaders and officers offer to themselves, as well as to the way in which they justify themselves to their members. But that notion has always been implicitly in conflict with the polity's double commitment to full employment and to an acceptable level of inflation. In this country, at any rate, the evidence of the last twenty years suggests that low inflation can be combined with full employment only if wage increases are restrained by an incomes policy of some kind. An incomes policy, whether statutory or voluntary, requires the consent of the trade unions. But the unions cannot give that consent for any length of time. They can — indeed, they have — signed frequent declarations of intent or social contracts, and at the moment of signing they have almost always meant what they said. Sooner or later, however, they have always discovered that the essence of a trade union is to engage in free collective bargaining; and that it is impossible to continue to do that while operating an incomes policy, or at any rate while operating any of the sorts of incomes policy which have been tried so far in this country.

The result is that the old alliance between the trade-union establishment and the right-wing Labour Party leadership — the alliance which came into being after the fall of the Labour Government of 1929–31 and which constituted the mainstay of the party from 1931 until the early seventies — fell apart.

The unions did not swing to the left as is sometimes said. They did, however, swing away from their old allies on the parliamentary right. They swung, not because they had suddenly been converted to, or even because they had slowly rediscovered the virtues of, some left-wing socialist ideology, but because their old right-wing allies were forced, by the exigencies of the economic situation, to abrogate free collective bargaining in the interests of price stability. But that did not make the swing any less pronounced, or any less portentous in its consequences. The balance of power in the Labour Party was changed fundamentally, and almost certainly permanently. In order to justify their behaviour, moreover — behaviour in which they were bound by their very nature to engage during the periods when incomes policies were breaking down — the union leaders became increasingly prone to appeal to the old 'zero-sum' notions of the class struggle. Thus, in 1973-4, when the National Union of Mineworkers was engaged in a singularly ruthless attempt to improve its differentials as against the rest of the working class, the leaders argued their case in terms that suggested that the miners were the vanguard of a heroic proletariat, engaged in the life-and-death struggle with the capitalist system; in the sort of language, in other words, which had been used fifty years before.

While the Labour Party was changing its character in this way, rather different changes were taking place in the Conservative Party. In an important sense, the Conservative Party of twenty-five years ago had no views about the use of power. It had views only about who ought to hold power. It stood for the double proposition that the traditional English governing class, which had run this country since the Glorious Revolution, should continue to do so; and that it was for the leaders of that class to respond, as seemed best to them, to whatever new problems they might from time to time confront. They might respond in what would conventionally be considered a right-wing way or in what would conventionally be considered a left-wing way; that hardly mattered. What mattered was that the response should be theirs: that their

class should be in charge. Between 1960 and 1980 – perhaps
even between 1964 and 1970 – that view lost its hold. A
Conservative Member of Parliament once told me that, soon
after his election in 1959, the then Conservative chief whip
took him on one side after a Division, looked in horror at
his feet, and exclaimed: 'You're wearing *suede* shoes!' It is
inconceivable that such an interview could take place today.
Nowadays, suede shoes are as common in the Conservative
Party as in any other party – metaphorically as well as
literally. The Conservative Party is still the party of the old
governing class. But the governing class no longer believes
in its own right to govern; and, even more significantly, non-
governing-class Conservatives no longer believe in its right to
govern either. Broadstairs and Grantham have triumphed over
Eton and the Guards. Since the party can no longer justify
itself as the party of a governing class, it has to find some
other reason for its existence and some other basis on which
to base its claim to power. Many of its members have found
it in an economic ideology – the ideology of the dissenters
of twenty-five years ago.

On a different, less obvious, but perhaps more important,
level the consensus of the fifties has been challenged by a
complicated change in the public mood: in the ethos, if not
of the whole of the society, then at any rate of its most
articulate and self-confident members. Everywhere, authority
is under question, under suspicion even, forced to justify and
to defend itself, in a sense which was not true thirty years
ago. Everywhere, the flag of revolt has been raised – not
perhaps by the entire public, but by significant and influential
sections of it – against the rather paternalist, rather centralist,
rather bureaucratic forms through which the consensus of
the fifties was put into practice. Industrial relations provide
an obvious example. One of the reasons why it has been so
difficult to operate an incomes policy for any length of time
is that the union leaders, with whom governments have
bargained, are no longer in control of their unions. In a quite
different sphere, the same mood has led Members of Parliament to demand measures of parliamentary reform involving

fundamental changes in the relationship between the House of Commons and the executive, the logic of which runs counter to some of the central assumptions of the 'Westminster Model' of the Forties and fifties — in particular to the assumption that ministers and ministers alone make policy and that civil servants cannot be held to account for policy decisions. Yet another manifestation of this mood is the growth of separatist movements in Scotland and Wales and the much fainter stirrings of regional sentiment in the north and south-west of England. A fourth is the growing tendency of groups which believe themselves to be disadvantaged or aggrieved to seek redress through *ad hoc* or single-issue protest organizations of one kind or another rather than through the political parties: to confront the system, in other words, instead of trying to use it.

Even more significant than these challenges to the old consensus is the response of its defenders. With the possible exception of Anthony Eden, the politicians who governed Britain in the forties and fifties, — Churchill, Butler, Macmillan, Lyttleton, Attlee, Bevin, Cripps, Morrison, Gaitskell — exuded an authority and a self-confidence which are hard even to remember today. The consensus was *their* consensus. They had fashioned it, and they had remade their country in its image. Not surprisingly, its virtues seemed self-evident to them; and they were, if anything, even more vigorous in its defence than its critics were in attack. The situation today could hardly be more different. Since the middle seventies, at any rate, the politicians of the consensus have seemed more and more uncertain of its validity, and their defence of it has been more and more faint-hearted. Mrs Thatcher owed her election as leader of the Conservative Party at least as much to the timidity and vacillation of her opponents as to her own — equally striking — courage and determination. Much the same was true of Micahel Foot's election as leader of the Labour Party. On both occasions, the pundits assumed that the consensus candidate would win. On both occasions, the pundits were wrong — not because they had not done their sums correctly, but because they had failed to reckon with the intangibles of zeal, enthusiasm, and adrenalin. The opponents of the old consensus knew what they were fighting for and loved what they knew. The defenders were half-hearted, unsure what they stood for, and half-persuaded that what they did not stand for might be right after all. In both parties, moreover, it has been the same story since the leadership election. On the central issues of economic management, Mrs Thatcher has routed the 'wets' in the present Cabinet, not because she has a better grasp of economic theory, but because she knows what she wants and they do not. On the central issues of party structure, the Bennite left of the Labour Party has won battle after battle, not because it is more numerous than the parliamentary right, but because it is more determined, more energetic, and more self-confident.

Once again, economic failure provides an important part of the explanation. The defenders of the old consensus know as well as everyone else that the promise of economic success, on which the whole edifice rested, has not been kept. Worse, they also know that it has not been kept because they failed to keep it. Worse even than that, they do not know why they failed: and, because they do not know, they cannot quite suppress the suspicion that the explanations so confidently offered by their opponents may have something in them. Thus, right-wing Labour ex-ministers cannot help wondering — to themselves, even if not to others — whether it might, after all, have been better to adopt the Tribune Group's 'alternative strategy' in 1976; thus, Conservative 'wets' half-accept the monetarist critique of Mr Heath's ill-fated dash for growth in 1972 and 1973.

But economic failure is only part of the story, and I doubt if it is the most important part. Many, perhaps most, of the politicians of the old consensus have been assailed by more fundamental doubts as well. Permeating all the different strands in that consensus was a robust — and, in retrospect, almost extraordinary — confidence in the capacity and efficacy of government as such. The administrative machine was seen as a machine and nothing more: and, moreover, as a machine which worked as it was supposed to work. Ministers would pull the levers; the wheels would move in the desired direction; and sooner or later the intended destination would be attained. Increases in social expenditure would buy the appropriate increases in social welfare. Comprehensive education would lower class barriers and promote social mobility. The amalgamation of local authorities would make local government more efficient. Industrial relations reform would cut down the number of working days lost in strikes. Confidence in government, moreover, was buttressed by confidence in the social scientists on whom government relied more and more heavily for expert advice. Anthony Crosland's blithe assumption that the 'modern economist's income analysis' made it possible — and would, for the foreseeable future, continue to make it possible — to maintain full employment was a particularly poignant example, but it was by no means the only one. Not only did economists know how to manage the economy: educationists knew how to improve the education system and sociologists how to abolish poverty in old age.

Half a generation later, little of that confidence remains. Even economists have lost faith in economics; among non-economists its prestige has fallen even more precipitously. The newer, humbler social sciences are in total disarray. The administrative machine looks less like a machine than a strange, clinging vegetable growth, impervious to the axe and the pruning hook. Ministers no longer pull levers, in the expectation that they will reach the intended destination. They are more likely to kneel down on the footplate and pray. Social engineering — and the old consensus was, above all, a consensus of would-be social engineers — has been discredited, not only in practice, but in conception and approach. Though the politicians of the old consensus no longer accept the intellectual and cultural premises on which it was based, however, they have nothing to put in their place. The dissenting creeds of the fifties have rushed in to fill the vacuum.

III

Both big parties, then, have been captured by the marginalized dissenters of twenty-five years ago. To be sure, they have not been captured in the same way, or to the same extent. Thanks to the social, economic, and intellectual changes discussed above, the Conservative dissenters were much stronger in the seventies than they had been in the fifties, but it is doubtful if they would have been able to capture their party if the Heath government had been more successful in its dealings with the trade unions. Mr Heath had been returned to Downing Street on a double ticket — no incomes policy, coupled with legislation to 'reform' industrial relations. Both halves of the ticket were abandoned. His reform of industrial relations law collapsed in the face of trade-union opposition; and he gave up his opposition to a formal incomes policy when he and his Chancellor decided to reflate the economy rather than allow unemployment to rise. His incomes policy, however, provoked further conflict with the unions; and in the terrible winter of 1973–74, he was systematically outmanoeuvred by the miners and, after a series of misjudgements and humiliations, removed from office by the voters. To anguished Conservatives, the moral seemed plain. Conservative-operated incomes policies would always provoke conflicts with the unions, irrespective of their intrinsic merits or demerits, because the unions did not accept a Conservative government's right to operate an incomes policy at all; and in a battle between a Conservative government trying to operate an incomes policy and a trade-union movement trying to destroy it, the electorate was as likely to side with the unions as with the government. The obvious conclusion was that, for Conservatives at any rate, incomes policies were a mug's game; and unless the Conservative Party were to abandon any hope of winning power again, that meant that it would have to find some other method of controlling inflation instead. The only other method on offer was control of the money supply.

The Conservative Party's conversion to monetarism was, in short, a response to the miners' victory in 1974. But although the monetarists argued that their case with great force and conviction, and although they managed to persuade their party colleagues of its validity, it rested on some distinctly shaky history. It was true that Mr Heath had lost his battle with the miners. It did not follow that his defeat had been inevitable. His government made a number of tactical errors, which a shrewder and more sensitive government would not have made; it also suffered a number of accidents, which a luckier government might have avoided. On a different and more important level, moreover, it is important to remember that the unions — including the miners — acquiesced in the first two stages of the Heath government's incomes policy; and that large numbers of workers acquiesced in the third stage as well. If the 1974 miners' strike was Mr Heath's Waterloo, it was, like the real Waterloo, a 'damn close-run thing': and the conclusions which eager Conservative monetarists drew from it went far beyond the evidence. Perhaps because of that, the Conservative Party's conversion to those conclusions was not as wholehearted as their proponents would have liked. The monetarist wing of the party succeeded in driving Mr Heath out of the party leadership, but its victory in the leadership election was more like a palace coup than a revolution. The old regime was defeated, but it was not destroyed. Survivors of it still hold high office under the new regime; and supporters are still plentiful on the back benches.

The Labour dissenters have won a much more decisive victory. Not only have they captured the party leadership, they have rewritten the party's policies and redesigned its constitution. Adherents of the consensus of the fifties still survive in the parliamentary party, but they have no support in the party outside parliament; and the power relationship between the parliamentary party and the party outside parliament has now been transformed, to the advantage of the latter. There is, of course, no way of telling what a Labour government would actually do if it were to win the next general election, but there is no longer much doubt that it would be pledged to unconditional withdrawal from

the European Community, to huge increases in public expenditure, to further nationalization, some of it without compen-
sation, and to a virtual siege economy. On a deeper and
more important level, moreover, the surviving Labour
adherents of the old consensus have been forced on to the
defensive much more completely than have their Conservative equivalents. The Conservative 'wets' can claim, with
some justification, that they are better Conservatives than
Mrs Thatcher or Sir Keith Joseph. They can invoke the
sacred names of Burke and Disraeli; they can point to a
long and honoured tradition of Tory democracy; they can
depict their monetarist rivals as alien intruders in a party
which has always ranked politics higher than economics,
and commonsense higher than ideology. In the Labour
Party, by contrast, the consensus politicians of the fifties
were the heretics, and the dissenters the keepers of the
true faith. It is Mr Foot and Mr Benn, not Mr Healey or Mr
Hattersley, who can invoke the party's sacred names: the
Manifesto Group, not the Left, which can plausibly be
depicted as alien. The consensus of the fifties, to put the
point another way around, went with the grain of the Conservative Party. It went against Labour's grain. Now that the
conditions which gave rise to the consensus have disappeared,
it is not surprising that Labour should have swung to the
left more sharply than the Conservatives have swung to the
right.

In the end, however, it is the similarities between the two
big parties that stand out, not the differences. Albeit to
varying degrees, both have repudiated the consensus of
the fifties, and both at least proclaim radical ideologies,
from which the consensus politicians of the fifties shrank
in alarm. By a curious paradox, however, these trends have
coincided with a progressive 'de-alignment' of the electorate,
which has become less and less firmly attached to the big
parties and more and more volatile in its voting behaviour.
Hence, of course, the emergence and growth of the S.D.P.,
and its astonishing showing in the opinion polls and by-
elections during its first few months. Whether it and its

Liberal allies can succeed, between them, in breaking the big parties' monopoly of power, and in establishing a three-party system in place of the two-party system which has prevailed in this country for nearly sixty years, remains to be seen. What is clear is that the S.D.P.–Liberal alliance has struck a chord which millions of ordinary voters find attractive; and it is at least possible that they find it attractive because they find consensus attractive as well. The consensus of the fifties has passed away, and cannot be resuscitated in the quite different conditions of the eighties. Apart from any other considerations, the constitutional arrangements on which it rested are in disarray, and the economic environment which sustained it has changed out of all recognition. But it does not follow that the harsh and clamorous dissension of the seventies is bound to continue. 'Consensus is dead', the voters seem to be saying. 'Long live consensus!' It is a message which the politicians will neglect at their peril.

Marxism and Communism

Wlodzimierz Brus

I wish to examine the relevance of Marxism as a social theory for the practice of communism. By the latter term I shall mean the system actually existing in countries ruled by Communist parties, in spite of their self-designation as 'socialist'. The cases in which the term 'communism' (or 'socialism' for that matter) will have to be used as a model designation will be duly indicated.

The communist systems are by no means identical in all the countries concerned. Particularly significant are the differences between the communisms of the Soviet bloc, Yugoslavia, and China, a fact which immediately brings out the largely independent revolutions in the two latter countries. I shall not dwell on these differences, save at some points when they may acquire particular importance for our discussion.

I do not pretend for a moment to be a social theorist in the strict sense. This, combined with the quite recent publication of Leszek Kolakowski's outstanding three volumes on *Main Currents of Marxism*,[1] explains my reluctance to sail out into the deep philosophical waters of social theory proper. I would prefer to describe myself rather as a *consumer* of social theory, especially in its application to political economy — the political economy of communism in the first place. Hence, as far as possible, the discussion will be kept within these boundaries.

I

The Marxism–communism plane seems particularly attractive as an exemplification of the general problem of the relationship

---

[1] Leszek Kolakowski, *Main Currents of Marxism*, 3 volumes, Oxford University Press, Oxford, 1978.

between social theory and political practice. Two features of the communist system account for this: (1) it has not evolved (at least, so far) within the confines of a preceding social order, but has arrived as an outcome of a conscious activity of a revolutionary state following a pre-conceived design (or what has been regarded as such); (2) the thus-established system is supposed to operate in a planned way, in terms of ex-ante selected goals, and in terms of ex-ante determined ways through a controlled process of allocation of human and material resources on a national scale. That is to say, both at birth and throughout the lifetime of communism the visible hand takes over from the invisible one. This singularity is expressed by the American political economist Charles Lindblom, in his attempt to analyse the communist system, as an 'imperfect approximation' to a model of an 'intellectually guided society' which assumes the ability of the leaders 'to produce a comprehensive theory of social change that serves to guide the society'.[2] It goes without saying that it is Marxism which is being generally regarded — by followers and opponents alike — as the brain of the visible hand: Marxist states, Marxist societies, Marxist economies — all these designations are in common usage, a rather unprecedented perception of the alleged link between an intellectual doctrine and institutional structures encompassing today over one-third of mankind.

The examination of the relationship between Marxism and communism consists habitually of attempts to verify this perception — starting with the question of Marxist inspiration for the ideology and strategy of the Communist parties before seizure of power, and ending with the Marxist conceptual foundations of the existing institutions and practice of their operation. In other words — how Marxist are the so-called 'Marxist' states, parties, societies, and economies? An affirmative answer opens up the prospect of evaluating the theory itself — by assessing the effects. There is no reason to dispute

---

[2] Charles E. Lindblom, *Politics and Markets: The World's Political–Economic Systems*, Basic Books, New York, 1977, p. 248.

this procedure, but I am not going to use it this time. Apart from other considerations, to follow this approach in any meaningful way would require a detailed critical analysis not only of an enormous variety of conclusions, but also of criteria of arriving at them. Two random examples: the Soviet military advance which, depending upon different criteria, may prove or disprove the superiority of the system; and the Soviet model of industrialization which is, as a rule, evaluated quite differently in the first and in the third or fourth worlds.

What is attempted here is, at least to some degree, something else: namely to consider the validity of Marxist social theory for an understanding of and hence for offering solutions to problems faced at present by the communist system. The existence of serious problems and difficulties cannot be disputed by anyone, regardless of the side taken with regard to communism. It seems also beyond reasonable doubt that problems and difficulties encountered by present-day communism are not of a momentary nature, and that they will not recede automatically simply with the passage of time; some of them seem rather to grow than to decline over time. How helpful, thus, can Marxism be for communism? What kind of guide to action can it provide? Remembering my own limitations mentioned at the outset, I shall try to answer the following question: what is the 'use-value' of Marxist political economy for communism?

The main economic headache for communism as we approach the end of the century is the marked decline of its developmental dynamism. The appearance of this phenomenon in all communist countries invalidates explanations based on national peculiarities, even though inter-country differences obviously exist. References to a sort of periodical breathing space will not do either — there are all the indications of a clear trend, again not without fluctuations around the trend-line. Even leaving aside the Soviet inter-war growth rates as exceptional, the fall in the tempo is truly dramatic: compared with the 1950s the combined average annual rate of growth in European Comecon member-states was less than half in the 1970s (falling from over 10% to less than 5%) and nothing, so far, indicates a reversal of the trend in the 1980s. The decline would be certainly yet more pronounced if not for the substantial inflow of external resources to practically all countries (except the USSR) in the last decade. Over the past five years the downward tendency accelerated, with indicators falling regularly from one year to another (6% in 1976, less than 2% in 1980). The Soviet Union — the oldest and largest communist country and the best-endowed in natural resources — shows the steepest and most consistent fall, almost undistorted by upward swings. In each consecutive quinquennium from 1950 (except 1966–70) the rate of growth has been lower, reaching no more than 4% annually in 1976–80 compared with over 11% in 1951–5.

I hope to be forgiven for statistical excesses (no more in sight), but this is an issue of considerable importance for our subject. At the first glance the figures mentioned (all coming from official sources) look quite comfortable against a Western background, particularly during the trough of recession. Moreover, the slowdown of growth could in itself be regarded as a positive phenomenon, if it were the outcome of a deliberate redirection of resources toward increases in standard of living or protection of the environement. Alas, this is hardly the case: deceleration seems to go strongly against the intentions of communist planners who invariably get results below and spending much above their plans. This means that indicators of efficiency (returns related to outlays) are much worse than simple output figures, with adverse impacts on living standards and sometimes truly disastrous environmental consequences (plunderous exploitation of national resources to compensate for diminishing returns). An important, however unexpected, result of *forced* reduction of growth is the proliferation of imbalances and bottlenecks in the economy, especially when attempts are undertaken to overcome the balance-of-payments difficulties by dramatic cuts in imports.

It is hardly appropriate to go here into detailed analysis of this remarkable weakening of the economic dynamism of communism. Needless to say, however, that it must be most worrying for the communist leadership — both for practical and ideological reasons. The latter touch upon some fundamental aspects of the claim of communism's superiority over capitalism which to a very considerable degree rest on communism's capacity to secure fast growth uninterrupted by crisis. For generations the steeply rising statistical curves served as the most popular and seemingly irresistible propaganda assets. Last but not least rapid growth was regarded as the paramount factor of transition to a higher stage of communism (communism proper — here, of course, in its model-designation) featuring abundance of goods and distribution according to needs. The still formally binding Programme of the CPSU, adopted in 1961 at the XXII Congress, even has a date: by the beginning of 1981 the Soviet Union was to enter the era of full communism. For some time already it has become evident that this programme and its accompanying twenty year plan, which, with typically Khrushchevian recklessness contained a whole string of concrete quantitative targets, is beyond reach and has become a source of acute embarrassment. Hence all over the Soviet bloc a curtain of silence has been drawn on this issue; with the 'transition-to-communism' formula quietly substituted by — attributed to Brezhnev — a new formula of 'mature socialist society'. The latter is hailed as the supreme achievement of Marxist–Leninist thought in recent decades. It certainly is very convenient, both because of its emphasis on stability instead of moves forward (a proper reflection of the prevailing mood of conservatism in domestic matters) and because of the avoidance of any specific commitments which may prove embarrassing in the future. (Some cynics say this is its main merit.) Whatever its suitability for the communist leadership, the new formula must be regarded as an admission of a major defeat, perhaps of historical dimensions.

This brief economic digression was meant to illustrate the broad nature and the profoundness of the difficulties encountered by communist practice. In conjunction with what is known about other and related conflicts — social, national, religious, etc. — brought time and again into the open by explosions like the recent Polish one, the need for a theoretical diagnosis of the ills is pressing. Is Marxist political economy *capable* of meeting these needs? The emphasis must be on the potential (or capabilities) because any judgement based on actual contributions would certainly be distorted by the consequences of political totalitarianism, which — let us hope — may not be fully effective in policing thought, but is certainly effective in policing its manifestations in public. Getting an answer to the question posed in this way must therefore be doubly difficult. However, *hic Rhodus — hic salta*. In order to stage the attempt we must stop for a brief discussion of the relevant fundamentals of Marxist theory.

II

Marxist political economy is a *social* discipline par excellence: its subject-matter in the strict sense is the social side of economic processes (the relations of production, in Marx's terminology) interacting with the technological side (forces of production). As rightly stressed by Sidney Hook[3] the role of the economic factor in the Marxist theory of development of society (historical materialism) can be understood properly only when the specific content of the notion 'economic' in Marxism is fully grasped: it encompasses the totality of *human* relations in the processes of production and distribution of goods and services. I cannot subscribe to the view rather widespread amongst economists and strongly stated by Peter Wiles, for example, that 'Marxism has no interest in economic choice or the distribution of scarce resources between competing ends'.[4] It is true, however, that in Marxist political economy the stress is laid not on the abstract logic of choice as such, but social situations which determine both this logic and the behaviour of economic agents. Thus, Marxism rejects

[3] Sidney Hook (ed.), *Marx and the Marxists: The Ambiguous Legacy*, D. Van Nostrand, Princeton, 1955.
[4] Peter J. Wiles, *The Political Economy of Communism*, Basil Blackwell, Oxford, 1962, p. 47.

the conventional assumptions about equality of all participants in market exchange, and endeavours to discover the foundations of relative *power* in the market, particularly as the buyers and sellers of labour are concerned. It is this relative power which exerts the dominant influence on the outcome of market processes, i.e. on the distribution of purchasing power and hence on the choices made by the consumer, who in this sense can hardly be regarded as sovereign. According to Marxist theory the ultimate determinant of this relative power has to be sought in the *ownership of means of production,* which is the paramount element of the relations of production. The type of ownership of means of production determines ultimately both the objective-function and the cost-function, in others words the premises of rationality in a given economic system. The mechanism of social–economic development, which constitutes the main object of interest for Marxism, is described in terms of the dialectical interaction between the forces and the relations of production. In perhaps the most often quoted passage from his works, Marx says:

In the social production of their life, men enter into definite relations of production which correspond to a definite stage of development of their material productive forces. The sum total of these relations of production constitutes the economic structure of society, the real foundation, on which rises a legal and political super-structure and to which correspond definite forms of social consciousness. . . . At a certain stage of their development, the material productive forces of society come into conflict with the existing relations of production, or — what is but a legal expression for the same thing — with the property relations within which they have been at work hitherto. From forms of development of the productive forces these relations turn into their fetters. Then begins an epoch of social revolution.[5]

The picture of the process here is, obviously, grossly over-simplified; nevertheless this brave summary reflects the fundamental idea of Marxist political economy and indicates at the same time what could be its function in political practice

---

[5] Karl Marx, Preface to *A Contribution to the Critique of Political Economy* in Vol. I of Marx, Engels, *Selected Works in Two Volumes,* Lawrence and Wishart, London, 1950, pp. 328-9.

broadly understood: it could provide, namely, tools for analysis of the essentials of the *conflict* between production relations and the exigencies of the development of productive forces. Such analysis in turn should help in finding ways of overcoming the conflict. However the intellectual aspect, the identification of the sources of conflict and ways of surmounting it, does not suffice by itself for securing progress, since interests of particular social groups are vested in a given economic system as well as in particular institutional structures within the system, whereas other social groups press for change. Hence the thesis about the key role played by *class struggle* in the process of historical development.

What has been presented so far could be called the *general layer* of the theory: in every mode of production the question of *social impediments* to economic development arise, and hence also the question of the sources of these impediments as well as of the conservative interests linked to them. Marxism has, however, another theoretical layer which could be called *specific*, related to the analysis of a particular mode of production, the *capitalist* one, and to the transition to communism derived from the analysis of capitalism. (Again, the term 'communism' designates here the model of a socio-economic order, socialism being in this context the first stage of that order.) The conclusion that the developmental tendencies of capitalism lead inexorably to communism supposedly justifies the label 'scientific socialism' used by political movements of orthodox Marxist inspiration. Marx's contribution to the analysis of capitalism is regarded by many non-Marxists[6] as most valuable without their necessarily accepting his political conclusions. On the other hand, the legitimacy of generalizing the results or even the conceptual framework of Marxist analysis of capitalism to the entire historical process has met with a great deal of scepticism growing over time. What matters most from our point of view,

---

[6] e.g. Joseph A. Schumpter, *Capitalism, Socialism, and Democracy,* Allen and Unwin, London, 5th edn., 1976; and Isaiah Berlin, *Karl Marx*, Oxford University Press, Oxford, 1952.

however, is not so much the general aspects and aspirations of the theory as the singularities of transition to communism. Let us turn again to the text cited before: 'The bourgeois relations of production are the last antagonistic form of the social process of production — antagonistic not in the sense of individual antagonism, but of one arising from social conditions of life of the individuals; at the same time the productive forces developed in the womb of bourgeois society create the material conditions for the solution of that antagonism. This social formation brings, therefore, the pre-history of human society to a close'.[7]

The reason for this fundamental difference is sought in the radically new way of resolving the conflict between forces of production and relations of production: instead of substituting one form of private property for another, transition to communism abolishes the private ownership of means of production as such. Social ownership brings to an end the division between the minority of owners and the majority of hired employees, thus opening the era of full emancipation (disalienation) of labour.

At this point we are faced with an apparent paradox: if this is the case, if the postulated transition from prehistory to history really brings about the expected consequences, then Marxist political economy, and by implication social theory in general, loses its cognitive functions. This was the corollary arrived at by many distinguished Marxists before the Soviet revolution and in its immediate aftermath, e.g. by Rudolf Hilferding, Nikolai Bukharin, and Rosa Luxemburg. The latter expressed the idea in a characteristically vivid and categorical manner: 'The last chapter of political economy as a science is the social revolution of the world proletariat.'[8]

There is no need to dwell on the detailed argumentation in favour of this position. If, however, one accepts the conclusion of what I called the specific layer of Marxism (the

---

[7] Marx, op. cit., p. 329.
[8] Rosa Luxemburg, *Einführung in die Nationalökonomie*, Berlin, 1915 Polish edn., Warsaw, 1959, p.86.

disappearance of social conflicts under communism) then —
keeping in mind the singularity of Marxist political economy
as social science par excellence — one has to admit that the
point about the end of such science with the advent of
communism seems to be well taken. Lenin was clearly off-
the-mark when he registered his opposition on the margins of
Bukharin's book *The Economics of the Transition Period*
thus: 'Wrong. Even in full communism the proportion
between $Iv + s$? and accumulation?' Neither Bukharin nor
any other protagonist of the view referred to above overlooked
the need for maintaining quantitative proportions amongst
various sectors of the economy or the relevance of account-
ancy, economic engineering, management science, etc. On the
contrary, they insisted on maximum efforts to develop all
these disciplines pertaining to 'scientific organization of
productive forces'. What they expected to become redundant
was political economy as a social science because of a lack of
social (as opposed to technical) obstacles to economic
development under communism. Another way of expressing
the same idea could be found in a much quoted and misinter-
preted statement by Engels about 'management of people'
being replaced by 'management of things'.

Thus, under the postulated conditions there is no place for
Marxism as a social theory: it will not have anything to offer
to practice, although particular points made by Marx may
remain relevant and stimulating such as, for instance, the
invoked proportion between the surplus produced in the
consumer goods sector and the demand by the accumulation
sector.

The 'end of political economy' position became eventually
another heresy in the USSR and other communist countries,
which is hardly surprising in view of the power elite's constant
need to legitimize its actions as Marxist and scientific. What
remained from Marxism in the straitjacket of communist
'conflictlessness' was however truly pathetic. Residuals of
previous Marxist lines of thinking could be pursued during
the so-called transition period when some elements of private
economy still co-existed with the public one. Soon, however,

the traditional orthodox Marxist political economy was left only with a (usually fictitious) distinction between the state and cooperative-ownership, and the 'remnants of the past in people's consciousness', fed incessantly by the capitalist environment. All the rest was an unending song of praise for the system which secured 'full correspondence between productive forces and productive relations'. The gulf between this 'theory' and reality widened so catastrophically with the passage of time that the concept of 'full correspondence' had to be amended. (Stalin himself demanded this — quite inexplicably — in his last writings.) The new formula allowed (at least at the socialist stage) for 'non-antagonistic' conflicts, which opened a narrow outlet for more pragmatic considerations, as, for example, the possibility of changes in the system of functioning of the economy (the so-called economic reforms). Genuine critical analysis remained, however, severely constrained by two provisions. Firstly, the 'non-antagonistic' nature of possible contradictions between productive forces and actual forms of production relations excluded conflicts of interest and hence activity of social forces. Contradictions were supposed to be resolved in time and harmoniously through appropriate decisions of the 'party', i.e. the leadership. Secondly, the appearance of these contradictions and the need for institutional change was as a rule caused by obsolescence only, when solutions appropriate for one period proved to be less suited for another. As aptly observed by Nuti[9] this approach is used whenever the need for an economic reform becomes officially recognized: extreme centralization of management is justified by the 'extensive stage' of economic development, and the need for change comes and is acknowledged with the advent of the 'intensive stage'. In effect each solution fully befits its time — the party is always right.

All this should not be treated as tantamount to disqualification of the entire socio-economic literature published

---

[9] D. M. Nuti, 'The Contradictions of Socialist Economies: A Marxist Interpretation', *The Socialist Register*, Merlin Press, London, 1979.

under official auspices in communist countries. Since the
second half of the 1950s and especially in more recent
years, a number of significant works in economics and
sociology appeared (it would be difficult to say the same about
politics). However, valuable contributions to economics
certainly belong in the first place to the subject described as
'organization of productive forces', or to highly technical
mathematical fields. (In sociology one would have to refer
to strictly empirical studies or discussion of sociological
techniques.) The authors of these studies in most cases declare
themselves as Marxists but both the problems tackled and the
methodological tools used are far removed from Marxist
social theory in the strict sense. Marxist language, or better
to say, jargon, is omnipresent but it serves usually as a
protective net. The Polish economist Oskar Lange, author
of the celebrated essay on market socialism, wrote in 1964
that 'the development of political economy of socialism was
limited hitherto mainly to the technical balance aspects of
the economy'.[10] No essential change has taken place since.
Insofar as it remains in the magic circle of communism's
immunity to socio-economic developmental conflicts, Marxism
stays sterile.

(This assessment does not apply to Yugoslav literature for
reasons which will become clear in the next section. Political
constraints operate, however, in Yugoslavia as well.)

Thus, the necessary (though not sufficient) conditions for
Marxism becoming useful under communism is the recognition
of conflicts, the roots of which stem from the system itself.
This may sound shocking when confronted with the conventional view of the relation between Marxism and communism,
but seems perfectly logical in the light of the premise of the
theory and in the light of its well-known destiny in the
communist countries.

[10] Oskar Lange, 'Ekonomia Polityezna', *Wielka Encyklopedia Powszechna*,
t.III, PWN, Warsaw, 1964, p. 334.

III

The question which arises now is: what is the contribution to practice of *critical* Marxism, once it is open to the possibility of conflicts within communist reality? The recognition of conflicts obviously does not in itself foreclose the correctness of analysing them in Marxist terms: at the same time it may undermine the coherence of the theory as such. The answer to the question posed is made additionally difficult by the proliferation of various critical currents applying (or attempting to apply) Marxist methodology to the examination of communism. More often than not these currents fight each other politically with truly sectarian zeal. Nonetheless, I shall try to put the field in order somehow and to review ideas which seem to be of importance for the political economy of communism.

The overall common feature, which justifies in a sense the treatment of all the currents in question as emanating from the same theoretical stock, is the rejection of the proposition that *socialization* of means of production has been duly fulfilled in communist countries. Within this general position two groups of attitudes can be distinguished.

The first, which will be discussed as briefly as possible in view of its rather minor impact on the political practice of the communist world so far, denies the validity of designating the countries ruled by communist parties as socialist ones. There are three variants of this position:

(1) The system existing in communist countries (China was exempted until recently) is a *state-capitalist* one. (The French economist Charles Bettelheim is the main protagonist of this view.[11])
(2) The system has developed into something peculiar, a new socio-economic formation in its own right neither capitalist nor socialist, in which, in Sweezy's words, 'the new ruling class derives its power not from ownership and/or control of

---

[11] C. Bettelheim and B. Chavance, 'Le Stalinisme en taut qu'idéologie du capitalisme d'Etat', in *Temps Moderne*, No. 394, mai 1979.

capital but from the unmediated control of the state and its multiform apparatuses of coercion'.[12] (Forty years ago Rudolf Hilferding referred to the controversy 'whether the economic system of the Soviet Union is "Capitalist" or "socialist"' as 'rather pointless';[13] the connotations of his article seem to have been, however, different from those of Sweezy's.)

(3) The system existing in communist countries is of a *transitory* nature in the sense that it has come into being in conditions of immaturity, which generate bureaucratization and internal conflicts. (This is the orthodox Trotskyite position developed more recently on the economic side by Ernest Mandel.[14])

Regardless of the differences, blown-up in polemical fervour, the theoretical gist of all these views is to point to the non-socialist nature of the relations of production in communist countries as the main cause of adverse phenomena in economic, social, and political life. Among the most conspicuous manifestations of the non-socialist character of the system is the maintenance, and even more so the strengthening, of the role of the market and material incentives. Elimination of these in favour of comprehensive planning by so-called direct methods belongs therefore to the paramount means of removing obstacles to economic progress. The views of the first group are often described as criticism of communism from the left; they are certainly orthodox in the sense that they derive conflicts from the non-socialist nature of the system, which in turn implies that they accept what we have called the *specific* layer of Marxist theory and consequently expect harmony in a system which would deserve the name of socialist or communist.

The abandonment of orthodoxy in the sense indicated

[12] Paul M. Sweezy, *Post-Revolutionary Society*, Monthly Review Press, New York, 1980, p. 147.

[13] Rudolf Hilferding, State Capitalism or Totalitarian Economy' in Hook (ed.), op. cit.

[14] Ernest Mandel, 'On the Nature of the Soviet State', *New Left Review*, No. 108, 1978.

above provides the *differentia specifica* for the second group, which comprises such otherwise dissimilar currents as Maoism, Titoism, and revisionism. All of these accept socio-economic conflicts under socialism as legitimate and indigenous, and even in cases when terms close to 'non-antagonistic contradictions' are used,[15] the interpretation is substantive, not verbal.

Maoism, insofar as it can be treated as a coherent doctrine, manifests perhaps the highest degree of distrust towards the transformation of property relations as the sufficient basis for disappearance of class divisions. Soviet-style takeover by the state of means of production is viewed as particularly conducive to generation and petrification of elites, gradually constituting themselves into a new class. But, as Kolakowski rightly remarked, Maoism looks with distrust at any kind of stabilization as a potential carrier of such a danger. Sweezy attributes (rather approvingly) to Maoism the following theoretical position: 'The notion that abolition of private property in the means of production ushers in an essentially classless society which, given a sufficient development of the forces of production, will evolve in a harmonious way toward communism – this notion is exploded once and for all. In its place we have a conception of socialism as a class-divided society like all that have preceded it, and one which has the potential to move forward or backward depending on the fortunes of a class struggle.'[16] This class struggle, at least in theory, consists of a relentless and continuous mobilization of the masses against the political, economic, and cultural establishment, against income inequality and market forms of allocation of resources (in this respect there is a great deal of convergence between Maoism and our first group). Regard for the economic consequences of this kind of action is slight, if it exists at all (the 'politics in command' principle). From the point of view of the political economy of communism – which cannot be indifferent to the question of

---

[15] Mao Tsetung, 'On the Correct Handling of Contradictions Among the People', in *Selected Readings,* Foreign Languages Press, Peking, 1971.

[16] Sweezy, op. cit., p. 95.

institutional premises for proper utilization of the potential
to develop the economy and increase welfare — Maoism has,
therefore, virtually nothing to offer. The disastrous effects
of the 'cultural revolution' in the economic sphere (not to
speak about other fields) seem to bear this out.

Titoism is by these standards much more constructive; it
continues in its own way the marxist tradition of the expected
conjunction of the emancipatory and the efficiency values of
communism. The fundamental theoretical premise of Titoism
(to be found in numerous official documents, but perhaps in
its most articulated and consistent form in the 1958 Ljubljana
Programme of the LCY[17]) is the concept of socialization of
means of production not as a single act, but as a continuous
process. The revolutionary takeover by the state opens only
the initial stage of this process and constitutes, at best, the
lowest form of social ownership (contrary to the Soviet doctrine
which regards state ownership as the highest form). This is
because — even under the assumption that the state genuinely
represents the working masses — ownership rights are vested
in a separate institutionalized entity which *hires* labour and
has at its disposal the surplus produced. The process of
socialization must become therefore a movement away from
state ownership towards direct ownership of means of production by the immediate producers themselves. This takes
the form of employees' self-management, which is expected
to erase gradually the division between the managing and the
managed, and in this sense create the objective foundations
for disalienation of labour along with the appropriate conditions for fuller utilization of the productive potential of
the society. This process cannot proceed but in a struggle
against bureaucratic vested interests; its progress should
mean a genuine 'withering away' of the state, the dominant
position of which gives way gradually to a self-managed
organization of a 'free association of producers'. (The Yugoslav doctrine evidently links up here with the decentralistic
interpretations of Marx's views, particularly those expressed

---

[17] *Program Saveza Komunista Jugoslavije*, Kultura, Beograd, 1958.

in connection with the Paris Commune.[18]) The economic foundation of the development of self-management must be provided by giving enterprises the right to make their own decisions; what amounts to discarding of the centralistic model of planning based on commands and allocation orders for producer goods. The economy has to operate under a largely self-regulating market mechanism, with the important difference that not private entrepreneurs but self-managed collective entities participate in the market process; these entities are obviously sensitive to market signals and stimuli and pass them on to their individual members. The principle of distribution 'according to work' acquires here a very broad meaning with income differentials becoming a function not only of individual work contribution but also of collective entrepreneurship.

Thus, *market socialism* emerges as an economic counterpart of the process of socialization in the self-management model — quite the opposite to the postulates of the first group of critical Marxists, and also of Maoism. Self-managed market socialism is supposed to combine the positive effects of competition (flexibility in adjusting output to demand, interest in innovations, etc.) with the positive effects of social integration in the economic sphere. The price to be paid for all this is the necessity to tolerate, both in the economy and in social life as a whole, conflicts engendered by the market, sometimes attenuated but sometimes even aggravated by the overall systemic framework. The most acute negative consequences are expected to be countered, through a social plan on a macroscale based on voluntary co-operation of self-managed entities. (Unfortunately, in this respect neither the theory nor the practice in Yugoslavia are specific enough.)

Revisionism has many common features withTitoism, and for the orthodox, who use the term as a kind of political denunciation and abuse, the distinction is immaterial because Titoism is regarded as a brand of revisionism. However, I apply the term here to the intellectual–political movement

[18] Marx, 'The Civil War in France' in *Selected Works,* op. cit.

developing in the communist countries of Eastern Europe since the mid-1950s (except Yugoslavia), to a considerable degree as a reaction to Stalinism and in opposition to the official doctrine. The lowest intensity of this current of Marxism is manifested in the USSR (apart from the historian Roy Medvedev I would be at a loss to name anybody); and the highest in Poland, Hungary, and Czechoslovakia, although periodically similar views have clearly surfaced in the GDR as well — Robert Havemann may be regarded as one of the last representatives of revisionism, whereas Rudolf Bahro[19] seems ideologically rather closer to our first group. The characteristic feature of revisionism, unlike Maoism and Titoism, is that it has never, so far, become the foundation of a state ideology; this applies to Czechoslovakia in 1968 as well, although the political programme of the Communist Party seemed then to develop along the lines derived from revisionist analysis. This circumstance has been undoubtedly beneficial for revisionism intellectually, but makes the task of its concise presentation more difficult as we are faced with a broad range of individual positions.

However, in an attempt at a sweeping generalization, one could say that revisionism shares with Titoism the concept of *socialization of means of production as a process*, which actually can never be completed and therefore can never achieve the state of conflictlessness, even in the economic sphere. The development of employees' self-management and its economic base of decentralization and utilization of the market mechanism are regarded also by revisionism as indispensable elements of the process. However, revisionist economists reject as a rule the extreme solutions of the market socialism model, defending the need for effective central planning on a macroscale in a modern economy.[20] But the essential difference between Titoism and revisionism is the latter's categorical insistence on *political pluralism*

---

[19] See Rudolf Bahro, *The Alternative in Eastern Europe*, NLB, London, 1978.
[20] Cf. Wlodzimierz Brus, *Socialist Ownership and Political Systems*, Routledge and Kegan Paul, London, 1975.

conceived as the opposite to totalitarianism, or (in order to avoid offending numerous Western political scientists fond of more blurred designations) to the 'monoarchy' in which the single-party state strives to dominate and control all mani-festations of public life. According to this approach, social-ization of means of production progresses not only with the advance of local self-management (enterprise, territorial), but in the first place with the progressive subordination of the decision-making centres to effective public scrutiny. This requires, obviously, the creation of institutional conditions for choosing and controlling those in government, for freedom of association and of speech, etc. In the revisionist analysis political pluralism is regarded as the key factor of efficiency of a planned economy, and the lack of it as the paramount manifestation of conflict between the relations of production and the exigencies of the development of the forces of production in the actual practice of communist countries. The emphasis on pluralism links East European revisionism with social-democratic traditions and with so-called euro-communism.

IV

To complete my survey it seems appropriate to touch upon the problem of the significance of critical Marxism for the dissident or opposition movements in Eastern Europe. In view of the scarcity of material and which, in addition, relates to only some of the countries in question (mainly Poland, in part also Hungary, Czechoslovakia, and GDR), this should not be taken as a product of systematic examination but rather as impressions.

The overall picture — hardly controversial — is that of a steep decline in Marxism's influence compared with the first post-war years. Substantial reasons for this have to be sought in the negative political experience, namely in the failure of several consecutive attempts at reform of the system along the lines envisaged by revisionism, which on the political side was often identified with expectations of reformist ideas winning over the party and its leadership. A string of defeats on the domestic front (Poland 1956-7, GDR, and in a sense — as far as political pluralization is concerned — Yugoslavia in the early 1970s), and particularly the crushing from outside of the 'Czechoslovakian Spring' in 1968, dealt a painful, some would say fatal, blow to the idea of endogenous reformability of the system, which was allegedly only deformed by Stalinism. Attempts at political activization based on radically orthodox Marxist concepts close to the ones referred to in our first group ('An open letter to the party' by Kuron and Modzelewski in 1964, similar tendencies in Hungary in early 1970s, Bahro and the 'Manifesto' of the Kommunistenbund in GDR) passed without tangible results. Marxism itself always stressed its link with political practice — 'The philosophers have only interpreted the world differently, the point is to change it.' The inertia of the system coupled with growing costs of the failure to change it could not but severely impair the attractiveness and credibility of the theory.

However, the reference to political practice does not explain everything. Critical Marxism, trying to face up to the challenge presented by communist reality and to work out methodological instruments for meaningful analysis, often willy-nilly undermines the coherence of the doctrine itself. This peculiar feedback effect applies both to the strict orthodoxy and to revisionism alike. A few examples:

If we are to accept the explanation of bureaucratic deformations by the immaturity of socio-economic conditions prevailing at the inception of communist systems, then we have to question Marx's assertion that 'no social order ever perishes before all the productive forces for which there is room in it have developed; and new higher relations of production never appear before the material conditions of their existence have matured in the womb of the old society itself![21] The answer given by Trotsky and Lenin that socioeconomic maturity had to be measured not for individual countries but for *world* capitalism as a whole (with capitalism's reign breaking-up in the *politically* weakest link) may have

---

[21] Marx, Preface to *A Contribution* . . ., op. cit.

had some plausibility at the time, but is clearly unsatisfactory sixty years since, when the alleged Marxian regularities seem to operate exactly in reverse, and when Russia — regarded as embarrassingly backward at the time of the 1917 revolution — looks now developed in comparison with other countries carrying out their own revolution of a similar type.

If we are to follow Paul Sweezy in stating that 'We do not need to rule out [*sic*!] the *possibility* of a post-revolutionary society being socialist in the Marxian sense . . ., but we do need to recognize that a proletarian revolution can give rise to a non-socialist society'[22] then we have to revise one of the tenets of Marxist historical determinism, namely that communism is a natural *successor* to capitalism. From some point of view this would constitute a theoretical breach even more serious than the rejection of the unilinearity of the historical process, e.g. by fuller recognition of the place played by the so-called 'Asiatic mode of production'.[23]

If, finally, we are to agree with revisionism (and with other currents of critical Marxism as well, including Maoism in its own way) that abolition of private property is in itself not tantamount to socialization; that public owner-ship does not automatically mean genuine popular control over means of production, the overcoming of alienation, destruction of all roots of social stratification etc.; then we have to look again at the concept of ownership of means of production as the fundamental element of the relations of production, and perhaps revise the relationship between the economic base and political superstructure. Political power and monopolization of the communication media in a broad sense could then emerge as independent factors of power relations in the society and their link with ownership of means of production may prove to be not unidirectional but of a feedback kind. Moreover, under this approach there would be no ground for an *a priori* rejection of the assertion

[22] Sweezy, op. cit., p. 137.
[23] Cf. K. A. Wittfogel, *Oriental Despotism*, Yale University Press, New Haven, 1957.

that the negative experiences of communism, both in the sphere of economic efficiency and in the sphere of human freedom, stem from the very fact of dominance of public ownership, and not merely from its peculiar features acquired under totalitarian conditions.

More issues of this type could probably be brought up. However, what has been said already seems to suffice for the realization that not only political failures, but also theoretical weaknesses bear responsibility for the decline of the role played by critical Marxism in shaping the intellectual face of the oppositionary movement in communist countries.

This is not to say that Marxism is completely absent in this movement. Side by side with people fully dissociated from Marxist philosophical attitudes, political views, and socio-economic conceptions, we can find among the intellectual opposition − often in the first rank − people who for many years were thinking and acting as Marxists. Some of them still highly value the cognitive powers of Marxism and regard the critique of communism in Marxist terms as particularly pertinent and fruitful − despite all the problems and reservations. Others, without labelling themselves Marxists any more, cultivate some traditionally Marxist approaches, e.g. to the role of the working class in social transformations. This is the point made by Bence and Kis, two Hungarian authors who not so long ago struggled for the renaissance of East European Marxism and later presented the reasons for their change of mind in an article tellingly entitled 'After the Break'.[24] Even Milovan Djilas detects some traces of Marxism in his intellectual stance: 'First I toppled Stalin, then I fixed my critical sights on Lenin, and ended up (though not in 1954) with a qualified rejection of Marx as well. I say "qualified" advisedly, because for a long time after 1954 I went on being a Marxist. The *New Class* is, in its basic approach, still a Marxist book although it contains many unorthodoxies. Even now I am sometimes under the

[24] G. Bence and J. Kis, 'After the Break' in F. Silnitzky (ed.), *Communism and Eastern Europe*, Harvester Press, Brighton, 1979.

impression that my mind is not completely free from the Marxist type of formulation.'[25]

Marxism, or elements of Marxism, in the independent intellectual life in communist countries is by and large free of the arrogance which was so pronounced in the utterances of its founding fathers, let alone of those of their followers who found themselves heading mass revolutions (which perhaps cannot be triggered off without the certainty of the faithful). Present-day Marxism in communist countries has discarded the unswerving conviction of its wholistic properties, of the absolute cohesion of its theoretical constructs, of its capacity to explain in full socio-economic processes, and hence to trace all the ways of action for political practice. For a social theory, aspiring to promote knowledge (I am sufficiently impressed by the 'Oxford terminology' to avoid the term 'social science'), this is a very healthy and promising development. However, because of the long and persistent pretensions to the stature of an oracle, the indispensable verifications and revisions have for Marxism today very grave — sometimes excessively grave — consequences.

[25] See George Urban, 'A Conversation with Milovan Djilas', *Encounter*, December 1979.

Index of Names

Adorno, T.W. 9n
Alembert, J.d' 40
Althusser, L. 77, 118n, 127
Arendt, H. 23, 67
Aristotle, 15, 67
Aron, R. 58
Attlee, C. 137, 149

Bahro, R. 174, 176
Ball, T. 23
Bence, G. 178
Benn, T. 122, 131-2, 154
Bentham, J. 39
Berlin, I, 164n
Bernstein, E. 45, 58
Bettelheim, C. 169
Bevan, A. 137-9, 142
Bevin, E. 149
Birch, N. 143
Bleicher, J. 6n
Boland, L. 99n
Bottomore, T. viii, 20
Brezhnev, L. 25, 161
Brus, W. ix, 22-3, 88, 174
Bukharin, N. 165, 166
Burke, E. 154

Calabresi, G. 160n
Churchill, W. 149
Comte, A. 4
Connolly, W. 118n, 125n, 132n
Cripps, S. 140, 149
Crosland, A. 137-8, 144, 151

Dahrendorf, R. viii, 20
Dallmayr, F. 18n
Descartes, R. 119
Diderot, D. 40
Disraeli, B. 154
Djilas, M. 178-9
Dunn, J. viii, 21-2, 117n, 122n, 124n,
    125n, 133n
Durkheim, E. 39, 45, 52-6

Easton, D. 71, 76
Eden, A. 149

Engels, F. 41, 166

Fay, B. 23
Feuerbach, L. 42
Foot, M. 88, 149, 154
Frankfurt school, 13, 30, 59
Friedman, M. 25, 99, 123

Gadamer, H-G. 6-9, 16-19
Gaitskell, H. 142-3, 149
Gay, P. 58n
Gramsci, A. 33, 39

Habermas, J. 13-21, 41, 67, 117
Halsey, A.H. 88
Hattersley, R. 154
Havemann, R. 174
Hayek, F.A. 123
Healey, D. 154
Heath, E. 150, 152-3
Hegel, G.W.F. 28-32, 35, 42
Heidegger, M. 31
Held, D. 13n, 117n
Hicks, J.R. 92
Hilferding, R. 45, 47, 165, 170
Hobbes, T. 114-15
Hobhouse, L.T. 50
Holland, S. 123
Hook, S. 162
Hubert, R. 40n
Hughes, H.S. 57-8
Hume, D. 112

Joseph, K. 122, 130, 154

Kant, I. 30
Keat, R. 23
Keynes, J.M. 39, 47
Kis, J. 178
Kolakowski, L. 58n, 157, 171
Konrád, G. 45
Kühne, K. 47n
Kuron, J. 176

Lange, O. 168
Laslett, P. 23

Lenin, V.I. 123, 166, 176
Lévi-Strauss, C. 40, 56
Lichtheim, G. 58n
Lindbeck, A. 92n
Lindblom, C. 158
Lipset, S.M. 23
Locke, J. 134
Lowith, K. 50
Lübbe, H. 36
Lukács, G. 40, 46n
Luxemburg, R. 165

Mao Tsetung, 171
McCarthy, T. A. 18n, 117n
Machiavelli, N. 67, 116
Macmillan, H. 137, 139, 143, 149
Malthus, R. 89
Man, H. de 43
Mandel, E. 170
Mannheim, K. 41
Marcuse, H. 45
Marquand, D. ix, 22
Marshall, A. 96
Marx, K. 25, 30-5, 40, 42-7, 50, 53,
    56-8, 64, 68, 77, 88, 89, 94,
    113-14, 162-5, 172-3, 176
Mauss, M. 52-3
Medvedev, R. 174
Mengelberg, K. 44n
Middlemas, K. 141
Mill, J. S. 4, 89
Mills, C.W. 45
Mitzman, A. 53n
Mommsen, W. 51
Montesquieu, C. 67
Morrison, H. 149
Myrdal, G. 88-91, 94-5, 101

Newton, I. 97-8
Nisbet, R. 40
Nuti, D.M. 167

Palmer, R.E. 6n
Parsons, T. 53
Pascal, B. 116n
Plato, 28-9, 65-6, 116, 132

Popper, K. 9-16, 19, 30, 35-6
Powell, E. 88, 143

Quine, W.V. 80

Rabinow, P. 6n
Reagan, R. 96
Ricardo, D. 89
Richter, M. 23
Rorty, R. 123n, 124n
Rousseau, J.J. 67, 72
Runciman, W.G. 23

Saint-Simon, H. 39
Samuelson, P. 99
Schumacher, E.F. 66
Schumpeter, J.A. 39, 45, 56, 164n
Sen, A. viii, 21, 98n, 99n
Simmel, G. 39, 56
Smith, A. 39
Sorel, G. 58
Stalin, J. 167
Stein, L. von 43, 44n
Strachey, L. 39
Strauss, D.F. 31
Sullivan, W. 6n
Sweezy, P. 169-71, 177
Szelenyi, I. 45

Taylor, C. viii, 20-1, 69, 93-4
Thatcher, M. 88, 96, 128-31, 145,
    145, 149, 150, 154
Thorneycroft, P. 143
Tocqueville, A. 67
Tönnies, F. 57
Trotsky, L. 176

Vico, G. 6

Weber, M. 17, 34-7, 39, 45, 46,
    48-51, 53-7
Wicksell, K. 94-6
Wiles, P. 162
Williams, B. 119n, 124n
Wittfogel, K. 177n
Wong, S. 99n